T0073986

TYPOGRAPHY FOR SCREEN

TYPE IN MOTION

HOAKI

Hoaki Books, S.L.
C/ Ausiàs March, 128
08013 Barcelona, Spain
T. 0034 935 952 283
F. 0034 932 654 883
info@hoaki.com
www.hoaki.com
hoakibooks

Typography for Screen
Type in Motion

Reprint: 2023, 2021

ISBN: 978-84-17084-13-4
Imprint: Flamant

Sponsored by Design 360° — Concept & Design Magazine
Edited, produced, book design, concept & art direction by
Sandu Publishing Co., Ltd.
info@sandupublishing.com
Chief Editor: Wang Shaoqiang
Executive Editor: Tan Jiahao
Designer: Liu Xian
Cover design and custom letters by Tauras Stalnionis
(www.tauras-s.com)

D.L. B 20841-2019
Printed in China

CONTENTS

PREFACE

Currently, there are two exciting and coinciding trends in the field of typography: an ever-increasing replacement of paper with digital screens on the one hand, and a steady flow of technological advances that is expanding fonts' functionalities on the other.

The first trend is pretty obvious. We each have digital devices around the house and in our pockets. And while we are strolling through a city, it is not possible to escape from omnipresent digital advertising. An important factor to keep in mind is that most of the content on these screens is viewed without audio. So in order to effectively get a message across, there is no way around using typography.

The second trend is a bit more under the radar for those who are less familiar with the latest developments in type design. It is best described by giving a few examples. Let us start with variable fonts. These are typefaces that break away from predefined weights. They are packaged in a format that allows end users to seamlessly choose any weight that is interpolated between a minimum and maximum value. Another big step forward is multicolour fonts. As the name suggests, these are fonts that are no longer limited to a single colour. Both multicolour fonts and variable fonts are getting more and more support from popular browsers. And another example is animated fonts—a field that I have specialised in myself. This is also a niche in the field of typography, in which fonts can stand out from the rest by using thoughtful movement to attract the eye.

Needless to say, digital screens offer a lot of possibilities compared to paper. And as algorithms get smarter each day, they open up an endless stream of innovative uses of typography, from mass customization to interactive typography in the form, for example, of type that adapts to a low-light environment for better readability or type that reacts to the weather, the time of day, or even the music playing on your phone.

It is interesting to ponder all the functional applications of these new possibilities. But it is equally (if not more) interesting to see how creative individuals explore these new boundaries in a more artistic way, without their being held back by things like readability or clients' notes. They can play around with type and technology in the same way as a carefree kid would play around with a box of Lego. In this way, they can also create and hack whatever makes sense in the moment and explore new terrain in the process. It might be whimsical or completely senseless at times, but I believe that this creative approach can give birth to new ideas that eventually find their way back into the functional and commercial realms.

I am excited to see where the collision of these two trends will take us. Will it cause an overload of irrelevant colours and motion in type? Or will it bring carefully crafted kinetic type into places where the use of it make sense? Anything is possible, and it is most likely to bring new levels of both horrible and brilliant typography to our attention. One thing is for sure: it will definitely change type designers' working processes. It will demand a more flexible approach within which they (might) have to take many more variables into account. And interpolation, environment, and animation are becoming their playing fields now.

The book you are holding explores a curated selection of innovative typography that is specifically designed for the digital screen. Each case is both beautiful and inspirational. But the fact remains that this is a book and not a screen. And even though a screen could never match the touch and smell of fresh paper, it can show animation and the RGB colour spectrum. So please do look up your favourites from this book in their natural habitat with sound, interactivity and all the bells and whistles. And when your device's battery is running low, pick up this book again and appreciate the nostalgic smell and touch of paper once more.

Jeroen Krielaars (Animography)

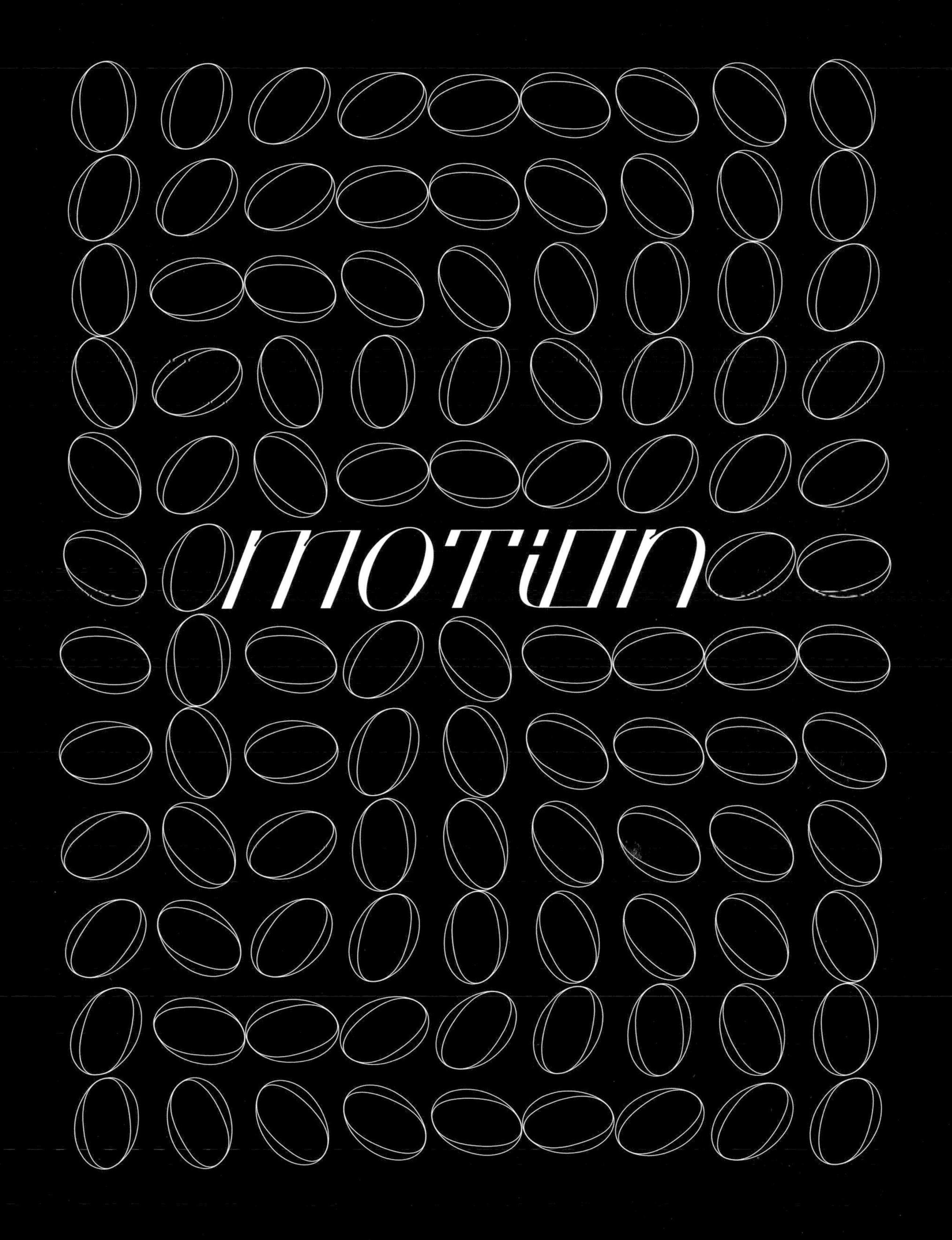
motion

Interview with ALEX FRUKTA

Alex Frukta was born in St. Petersburg, Russia in 1992. He has worked as a designer since 2011 and specialises in motion, illustration, and typography. He is the founder of Fontfirma Typography Platform, the co-founder of NORD Collective, and a member of Designcollector Team.

Can you tell us how did you come across motion graphics design?

When I created the static design, namely collage art and font design, I realised that I wanted to design something more complex regarding my projects. And motion design is a good choice for me. My story of acquaintance with the industry of motion design started logically. Besides, the logo rebranding of MTV in 2009 was also an incentive for me. As we know, the complete letter "M" was a key visual in the original logo. After the awesome rebranding, the "M" was horizontally cropped and gave the logo a wider proportion in a ratio of six by four. As I had been already interested in design, I was longing for something like that. Actually, it influenced and pushed me to make a decision of becoming a motion graphic designer. At the same time, I was also impressed by Vimeo as a cradle of motion design culture. I watched various works from other people and studios with great interest and desired to create something different.

You were born in St. Petersburg in 1992 and have worked as a designer since 2011. How do the Russian culture (even Soviet culture) and other cultures influence your design?

Right, I was born in 1992 in this beautiful and historical town of Russia. It is often considered Russia's cultural capital for almost 200 years. I was born in the Post-Soviet era, and the original name of this city—St. Petersburg returned. The city may affect me, but most likely, it is just a temporary foothold for me. The surrounding and bundle of moods have shaped me as a person who is engaged in motion design. Like a child playing with the Lego constructor, I discover new cultures with great interest and use them in my works combining with my vision and established traditions.

What is your workflow like? How did you communicate with your clients and prepare for your preliminary work?

Processes of communication are different from each other. They depend on a lot of factors. The client often plays a big role in the project. And agencies, managers, freelancers, brands, or companies can be regarded as my clients. I almost always have the same business arrangement with my clients: Firstly, I ask the client for a brief of the project; secondly, after my comprehension I meet with the client to discuss details such as scenario, timeline, design, guideline, budget, technical specifications, and others depending on the characteristic of the project.

What are the main challenges during your design of motion graphics? How did you solve them?

In the process of each project, there are moments when you have to invent and look for solutions. The only cure is the graphics applications: Adobe After Effects, Adobe Illustrator, and Photoshop.

The motion graphics you designed, such as "J&J 2018" and "G8 2018," have very unique animation typographic effects. How do you avoid some homogenised or cliché effects when creating motion graphics?

I am sure that clichés should not be avoided. It regarded as a tool for creating fantasy, and then everything will be spectacular. We are living in a continuously changing world of postmodernism. When a new stuff was created and attracting wide attention, it will always be reconstructed and redone into something else.

We have noticed that many of your works consist of different languages and characters, such as English, Russian, and Japanese. How do you break the boundary of languages and cultures to create resonant works?

Thanks for your assessment of my works. It is difficult to say how I combine those elements. It is always an unpredictable search with an unexpected outcome. During my creations, I like designing shapes and arranging letters. Such a process is fascinating and spellbinding. I can spend hours designing a single letter until it is up to the ideal effect. After some time, when I look back on the result I have achieved, I can say that I am proud of my work.

In recent years many typography designers have spontaneously started various projects of variable typefaces on Behance, Instagram, and so on. What do you think about this trend? Will such a trend affect your design style?

Just like the packages in the store or a cool design book like the one in which my projects are included, I will see and skip everything through the standard of whether I like it or not, and then it has already affected on me. Thanks to the cameras in smart phones, everything becomes simple—you can take a photo of what you like and do not need to memorise or sketch in the Moleskine.

Could you share your upcoming projects with us?

It is a commercial secret, but this is a good question, seriously. Usually, I work on different projects simultaneously and perform different tasks where design and motion apply in one way or other.

J&J 2018 (P012)

J&J 2018

BY Alex Frukta

According to the scenario of Johnson & Johnson (J&J), Alex Frukta had to stage five scenes that were regarded as the company's manifesto, achievements, and future plans. The pink colour used in the video is Alex's idea. This is a customized version only for his portfolio that shows a more attractive shade. The original one is in red for the company's corporate identity. Technically, Alex used Adobe Illustrator, Adobe After Effects, and Adobe Animate to draw and sketch the image of the future theme. And he usually draws the frame-by-frame animation on the tablet. To solve the difficulty of resolution which was more than 5K, Alex split the project into five animated scenes in Adobe After Effects according to the entire structure. Finally, Alex finished all five scenes and combined them into a common sequence.

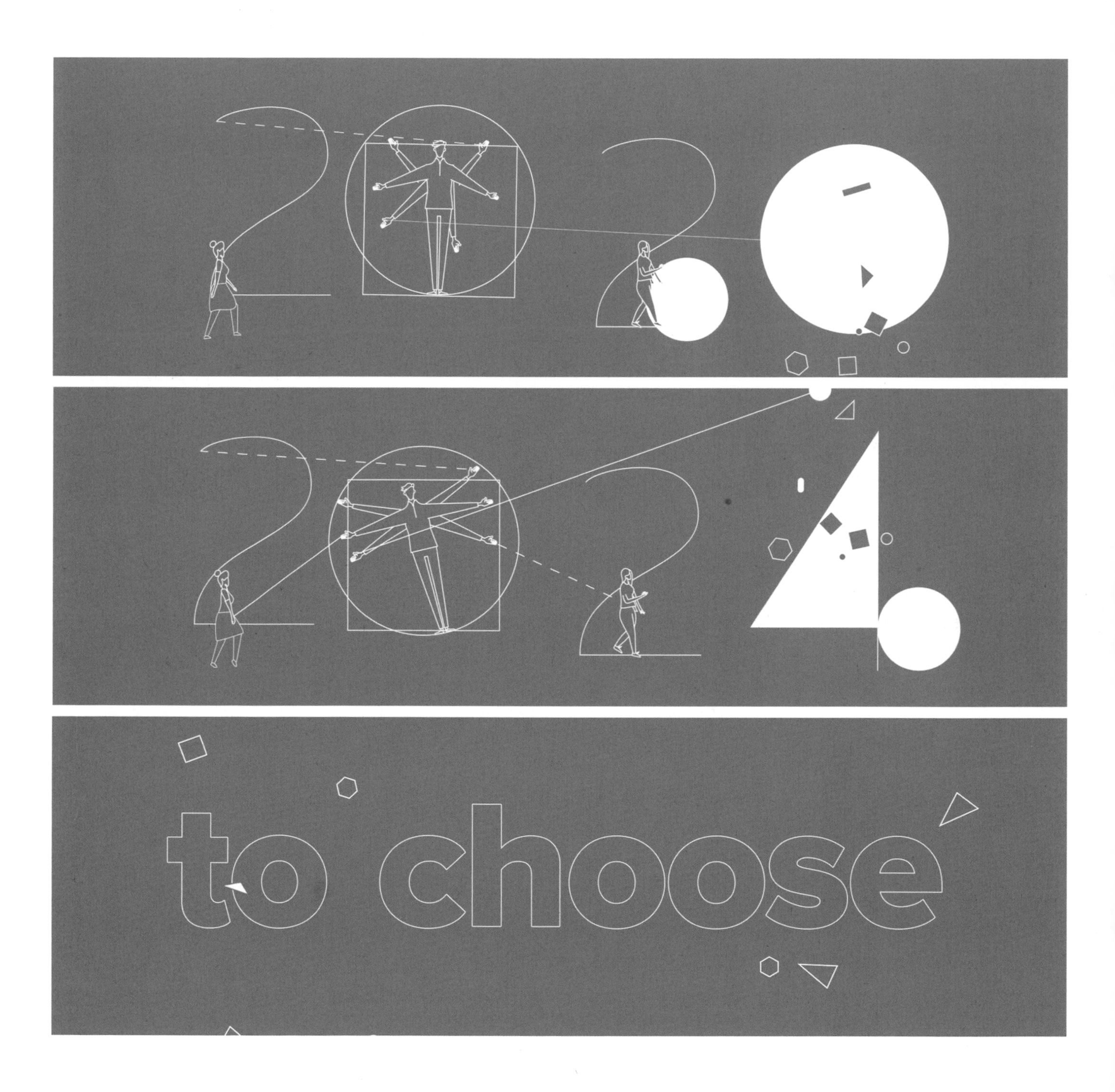

AGENCY **Illuminarium 3000** / MUSIC & SOUND **SoundDesigner.PRO** / CLIENT **Johnson & Johnson**

Behance
Vimeo

hidden and important
to see
to create
a new era

Grafika Awards Opening Video

BY Gimmick Studio

This video was created for the purpose of kicking off the 2018 Grafika Award Ceremony (a graphic design award). The animation can be described as a joyful explosion of typographic chaos. The project started with a very simple graphic identity then Gimmick Studio built a crazy typographic universe with 2D and 3D animations mixing and morphing in funny chaos.

PRODUCTION & MOTION DESIGN **Gimmick Studio** / GRAPHIC DESIGN **byHAUS, Gimmick Studio**
AUDIO **Cult Nation** / CLIENT **Infopresse**

Behance Vimeo

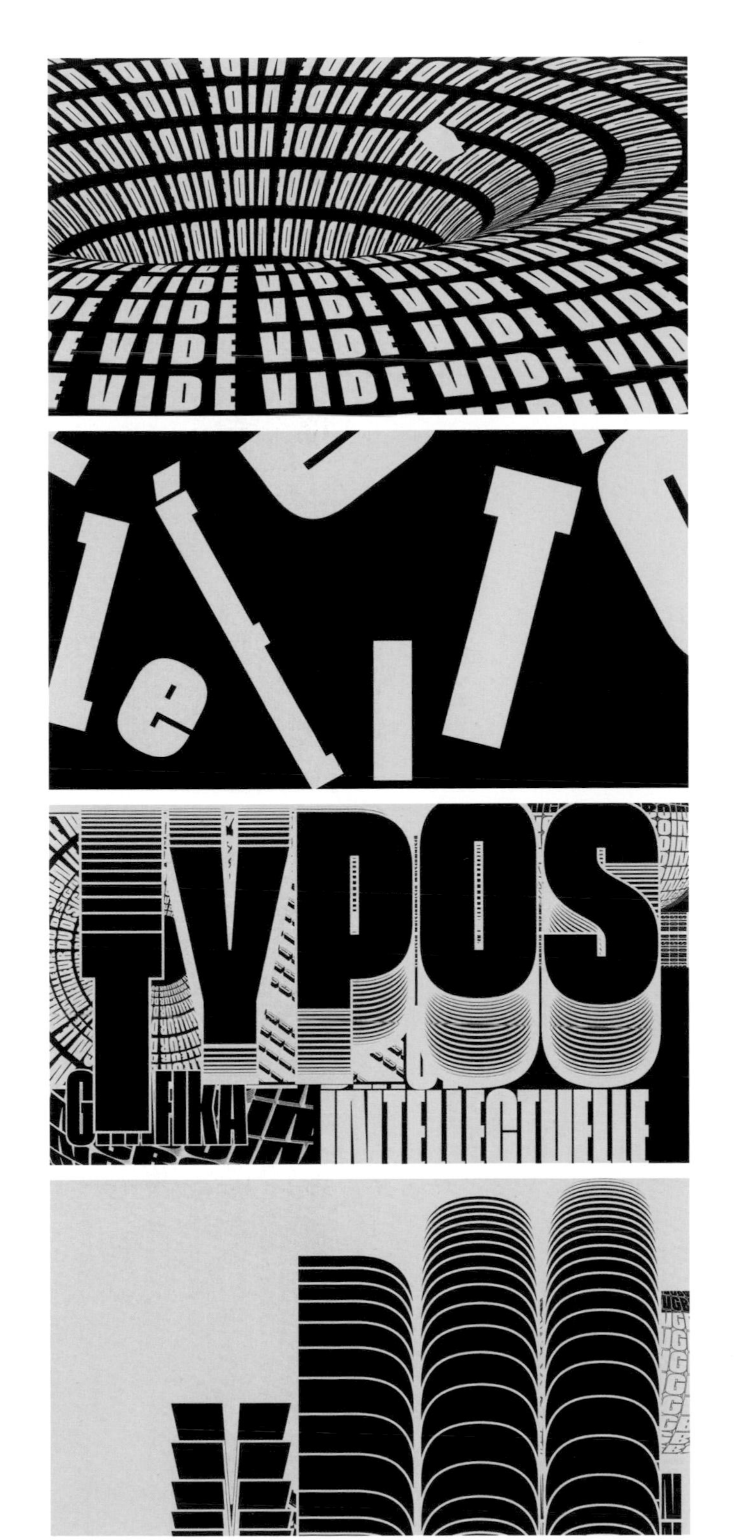

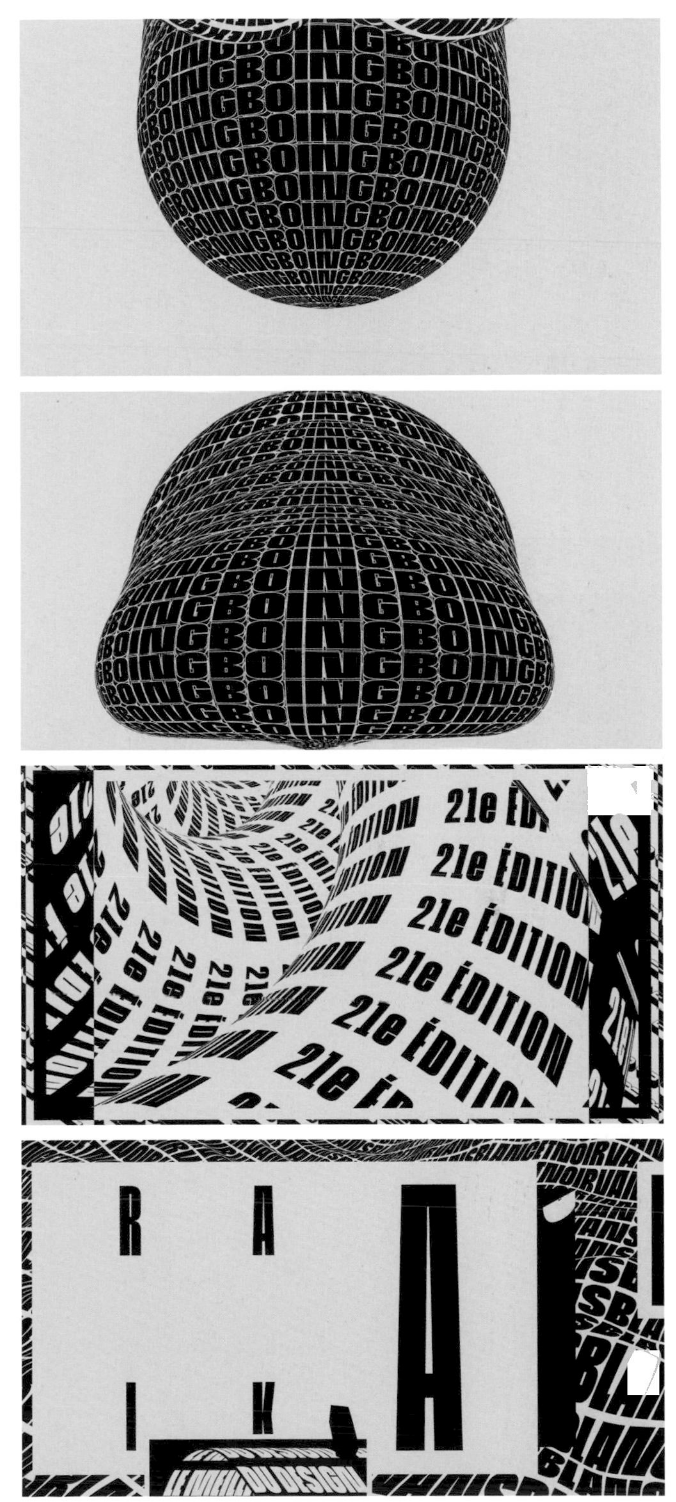

Arte Xenius

BY Éléonore Sabaté

Basing on the scientific fact that the white colour contains all other colours, Éléonore Sabaté started to animate the "Xenius" letters. Various colours are emerging from black and white moving letters. Each letter has its own principle of motion and distortion. Éléonore tried to illustrate more natural moves for each letter mechanically and structurally. For example, the shape of "e" is moving like a fluid and wave; the "i" keeps revolving on its own axis. Éléonore also edited the whole title by assembling the letters and playing visually with cuts and alternating the background from black to white with a positive and negative effect. Meanwhile, people can clearly see the title "Xenius" in white on a black background.

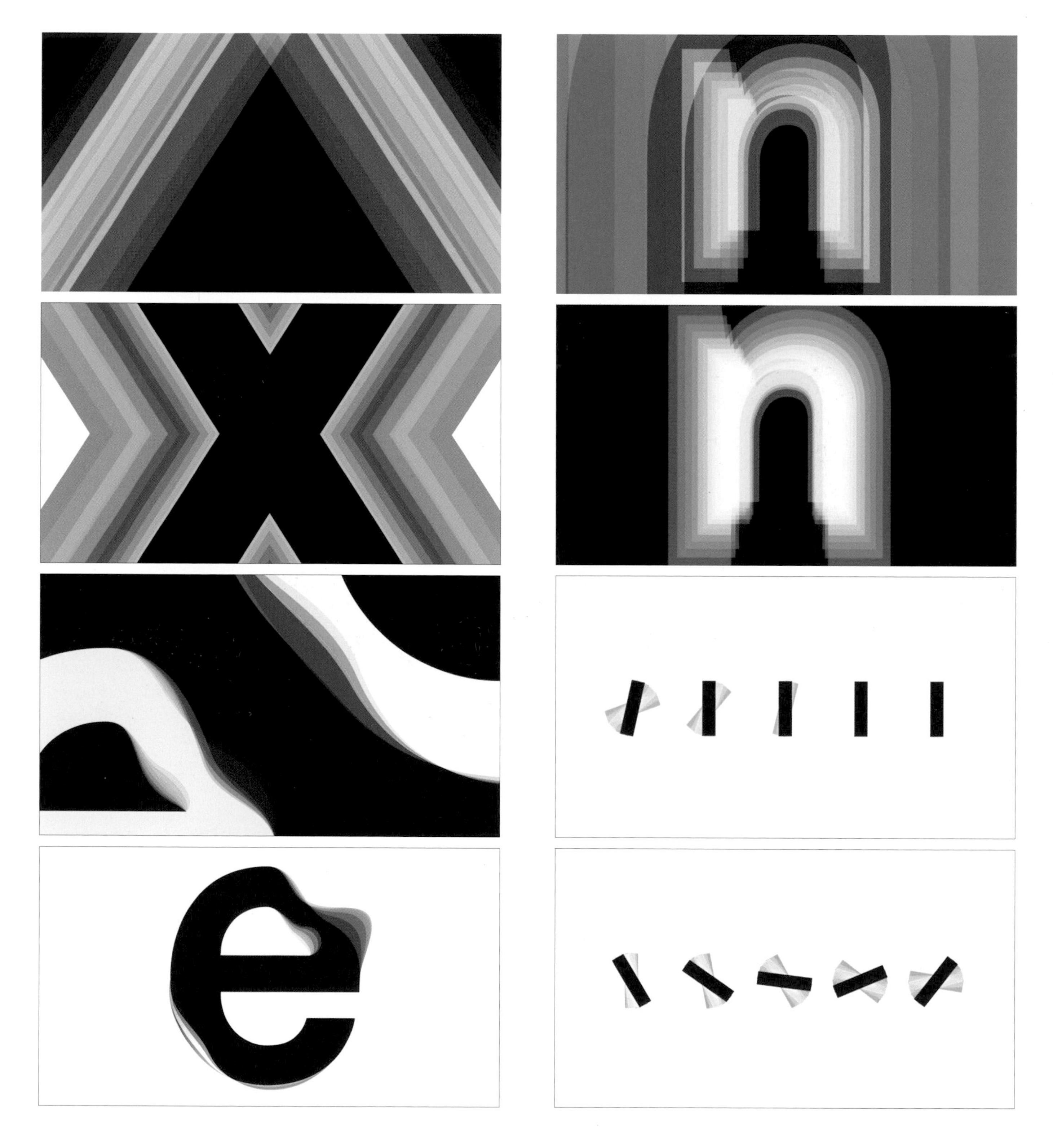

ART DIRECTION & MOTION DESIGN **Éléonore Sabaté** / COOPERATION & CLIENT **Arte**

Behance
Vimeo

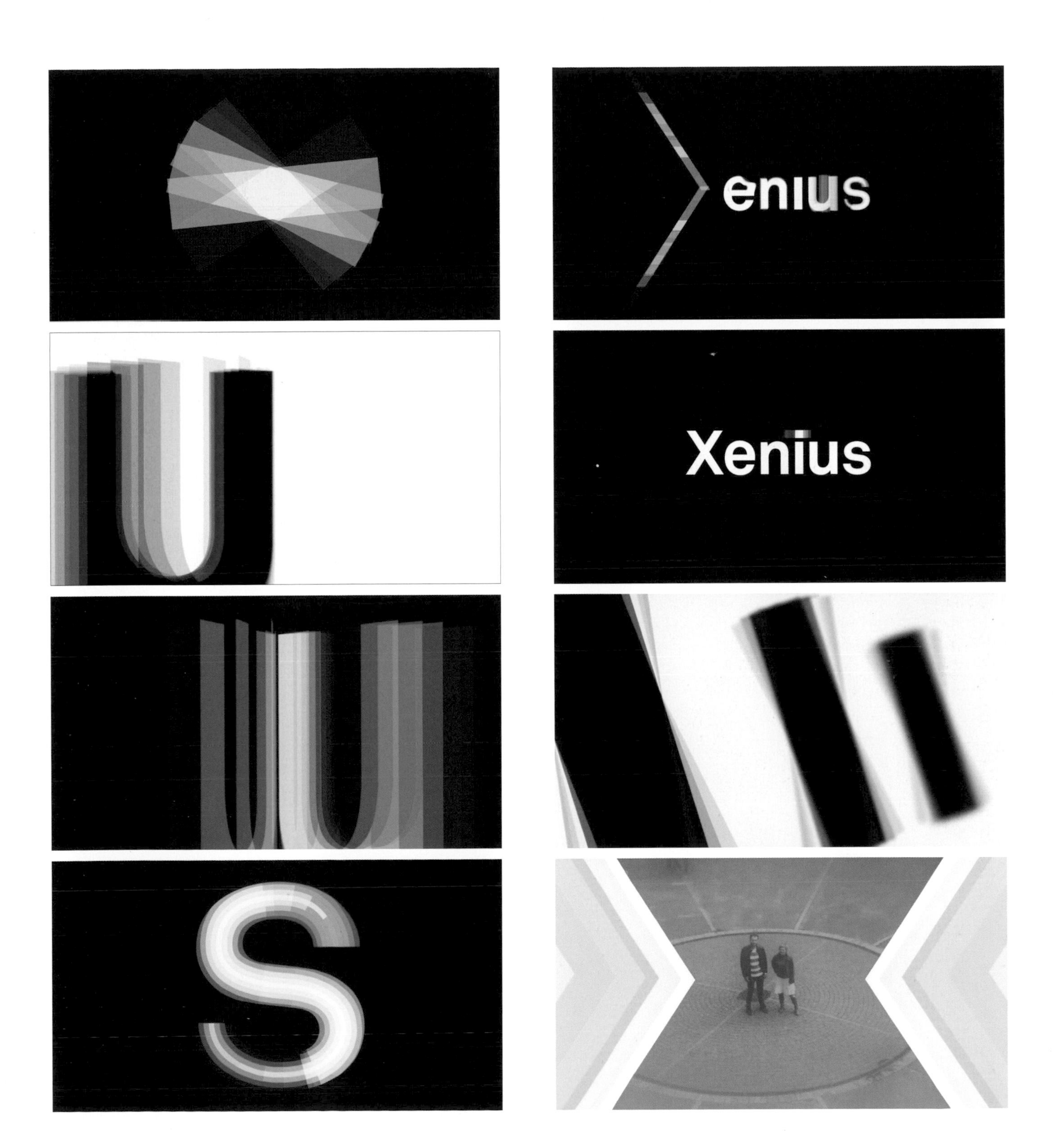
enius
Xenius

Nike Custom Typeface Oslo

BY Oh Yeah Studio

Oslo is a new custom typeface designed for Nike's concept store in Oslo, Norway. Nike wanted to let a local artist make a narrative thread through the new concept store. So Hans Christian Øren's brief was to create a bespoke typeface for the word "Oslo" and the sentence "Your only limit is you." Hans used Futura Bold as a reference. This video is an abstract audio-visual journey through a race from beginning to the end. By using the design elements from the typeface Oslo, such as the running track in Bislett Stadium, the video covers a physical outer battle with a decision to win the race, as well as a mental inner struggle against losing focus.

ABCDEFGHIJKLMNOP
QRSTUVWXYZÆØÅ
0123456789

GRAPHIC DESIGN & TYPOGRAPHY DESIGN **Hans Christian Øren, Oh Yeah Studio, Bekk**
MOTION DESIGN **Babusjka** / CLIENT **Nike**

Behance
Official

YOUR ONLY
LIMIT
IS YOU

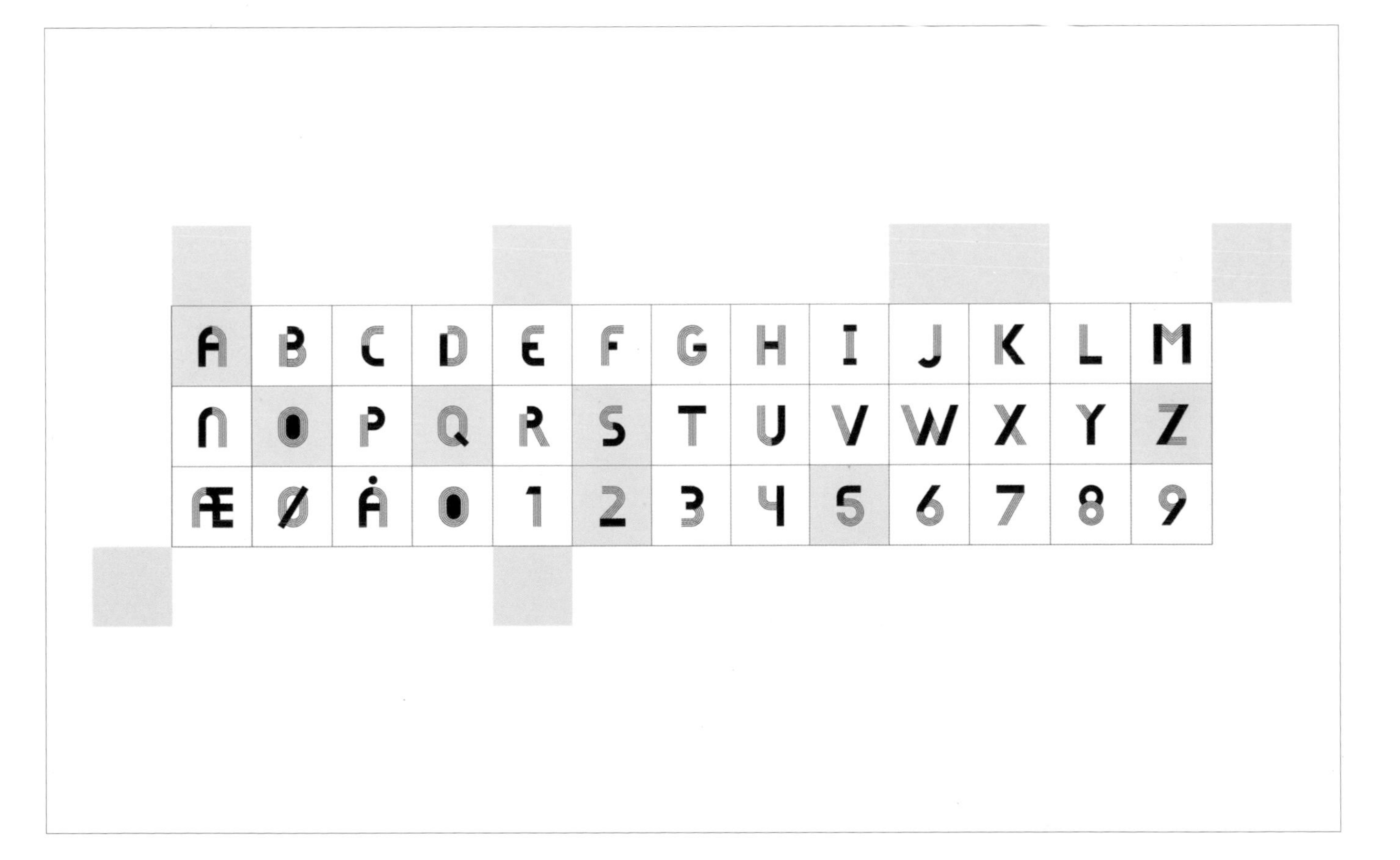
A B C D E F G H I J K L M
N O P Q R S T U V W X Y Z
Æ Ø Å 0 1 2 3 4 5 6 7 8 9

©2018— GLARE, LLC

BY Glare

Glare is a future-forward creative studio that launched in 2018. Founded by the digital artist duo Anthony Gargasz and Rafael Ramirez, Glare represents the new breed of creative consciousness that is constantly evolving with the resources of technology, music, and art. This video embodies the essence of Glare's creative vision, visceral style, and intentions of pushing the boundaries of digital art forward.

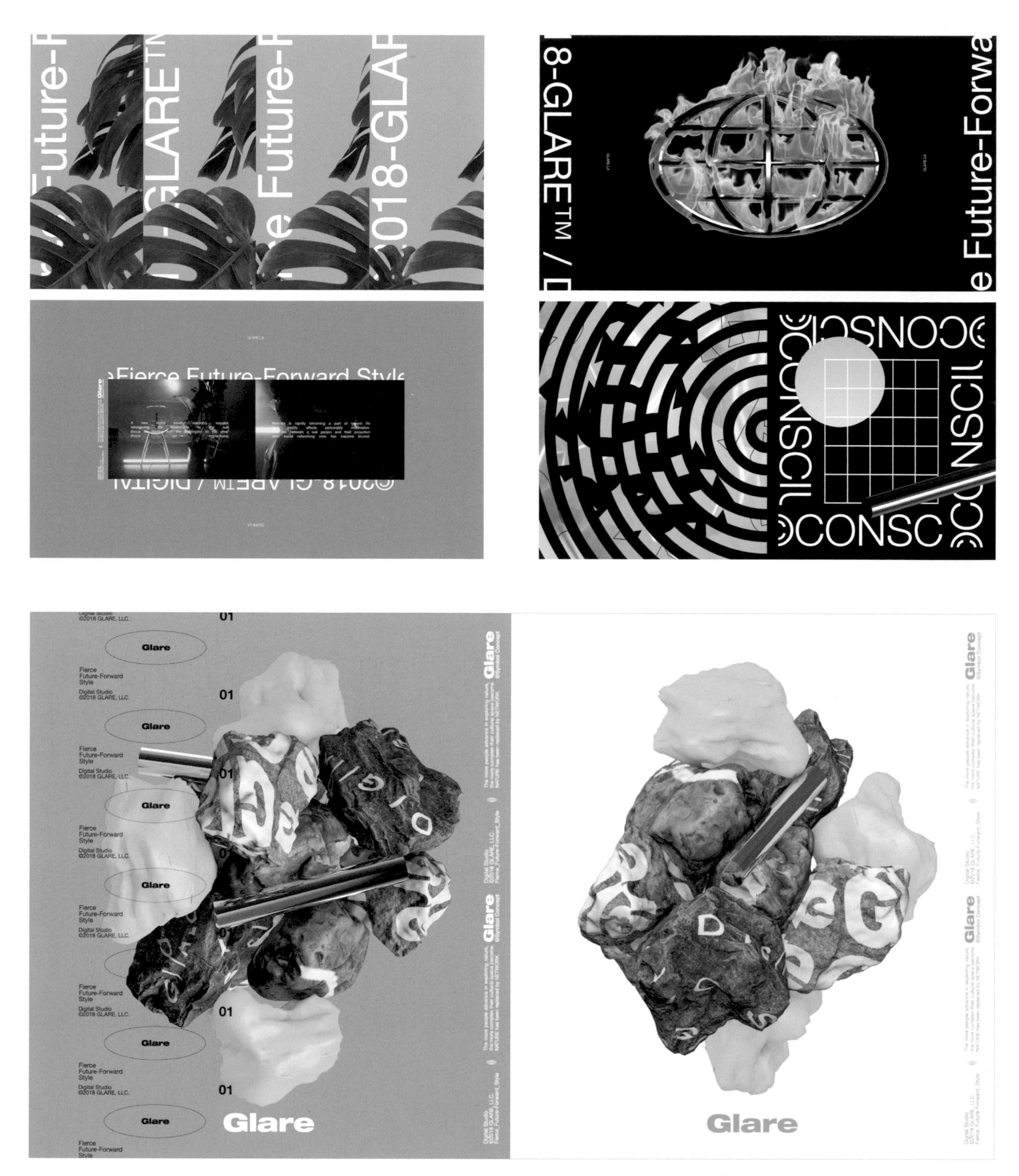

DIRECTION & PRODUCTION **Glare** / 3D MODELING **Dobu Haishen**
CINEMATOGRAPHY **Hunter Ney, Nina Hawkins** / PHOTOGRAPHY **Calvin Ma**

Behance
Official

Glare

©CONSCIUSNESS
©DIGITAL
©DIGITAL
©CONSCIUSNESS
Glare
Digital Studio
©2018 GLARE, LLC.
Fierce
Future-Forward
Style

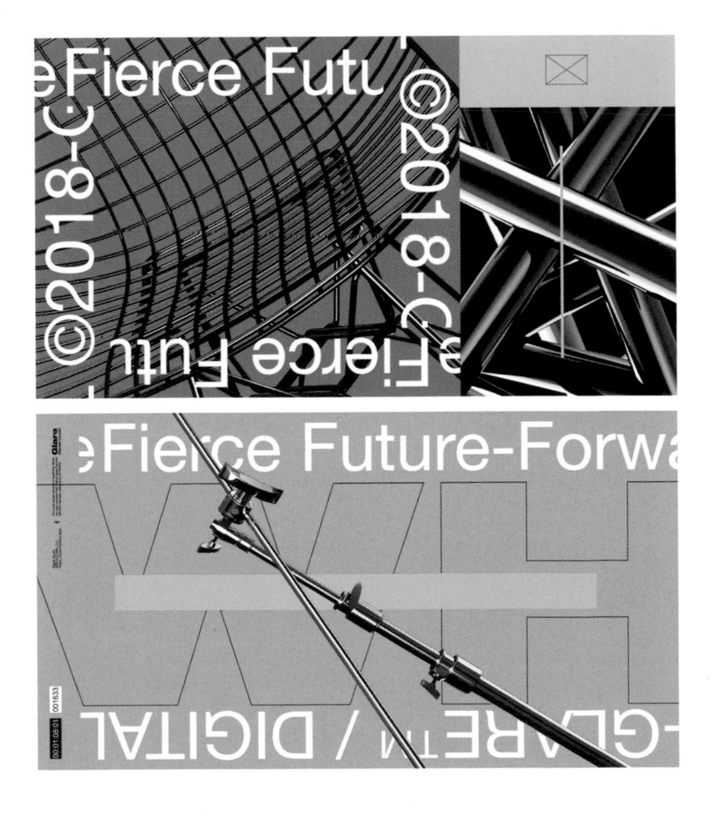
Fierce Futu
©2018-C
©2018-C
Fierce Futu
Fierce Future-Forwa
GLARE™ / DIGITAL

Adidas
N3XT L3V3L

BY République Studio

To celebrate the launch of Adidas' shoes N3XT L3V3L, République Studio took charge of the entire art direction of the event and also created a kinetic typography video displayed on a huge 20-metre screen. République Studio designed this one-minute video only with typography and colourful graphics. They expressed the idea of the N3XT L3V3L show in two sentences. Besides, they used only one typeface in this video, which is Next by Ludovic Balland. It matches perfectly the purpose that République Studio wants to show and of course the name of the shoes.

CLIENT **Adidas**

Official

BELIEVE IN Y URSELF
BELIEVE IN Y URSELF
BELIEVE IN Y URSELF
BELIEVE IN Y URSELF
BELIEVE IN Y URSELF
BELIEVE IN Y URSELF
BELIEVE IN Y URSELF
ADIDAS
N3XT
L3V3L
TEAM NDJOLI
TEAM NDJOLI
TEAM NDJOLI
TEAM NDJOLI
TEAM NDJOLI

BBC Radio 1

BY Mother Design

BBC Radio 1 has been the vanguard of the young music scene for the last 50 years in the UK. It needed a clear branding system as well as a contemporary new graphic language. The challenge for Mother Design was to develop an impactful and youthful brand, while also ensuring it can adapt to the needs of all media executions at the BBC. Mother Design also developed a suite of dynamic/punchy animations for video content and live events. These stings feature the Radio 1 logo constantly in movement, reflecting the freedom and dynamism of the brand.

CLIENT **BBC Radio 1**

Behance
Official

1

G8 2018

BY Alex Frukta

G8 is a festival of creative industries, which was created by RedKeds. Alex Frukta was invited to design the video for the festival in 2018. RedKeds asked Alex to avoid the video style of glitch and digital error, popular in motion design for the last five years. Alex got inspiration from the Soviet mechanical projector in which the filmstrips can show the images and plots step by step to the viewers. So Alex extracted some elements from the filmstrips for the key visual. Meanwhile, Alex showed the images with ten pattern backgrounds with different visual solutions of the figure "8" which are changing one by one. And the dynamic motion of the G8 logo's superposition is also embedded in the foreground.

MUSIC & SOUND **SoundDesigner.PRO** / CLIENT **RedKeds**

Behance
Vimeo

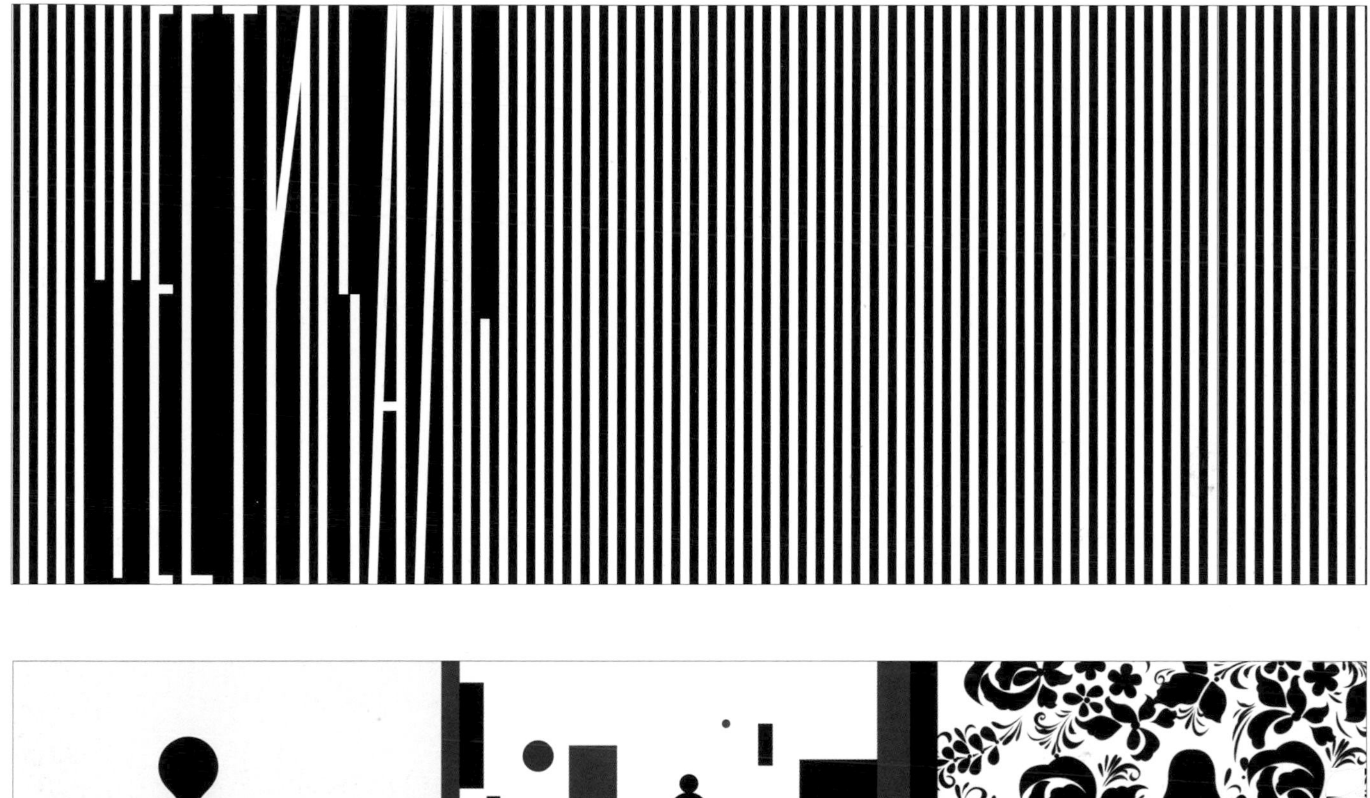

ZEN, KOMAZAWA, 1592

BY ALLd.

In this project, countless geometric patterns and words are drawn toward the Chinese character "ZEN" with a gentle movement like the flow of air, wind, or water. Such a movement with tranquillity carries the force like the gravity of the universe. It is an expression of human's motion, pulling the viewers toward "ZEN" and freeing them from the burden of the egos and worldly desire.

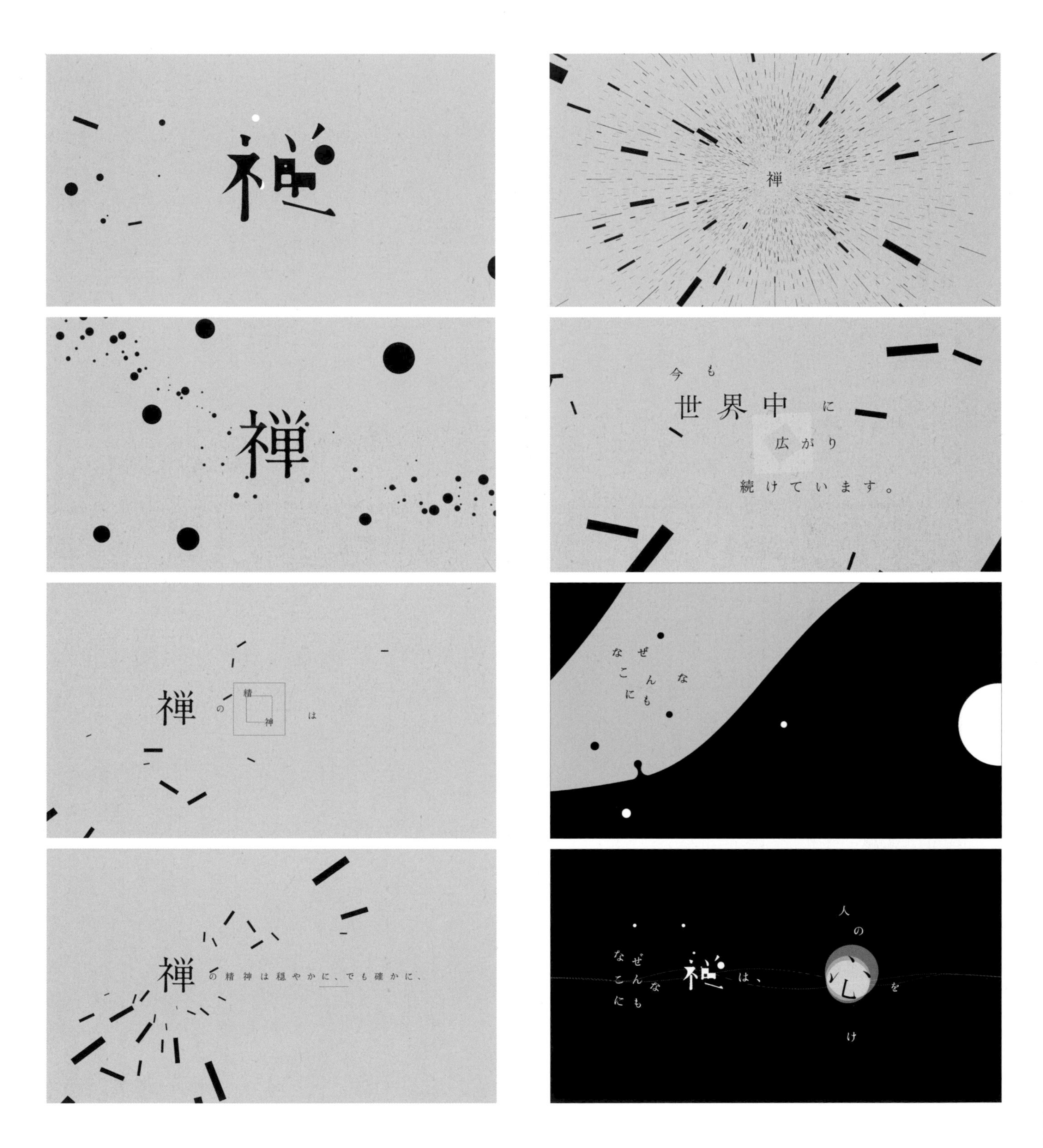

PRODUCTION **Atsushi Hashikura (northshore)** / DIRECTION **Kenichi Ogino (ALLd.)**
MOTION DESIGN **Kaito Mizuno (ALLd.), Masakazu Nomura** / DESIGN **Fumiya Hirose (ALLd.)**
SOUND **Satoshi Murai** / MANAGEMENT **Koya Matsunaga (northshore)**

Official

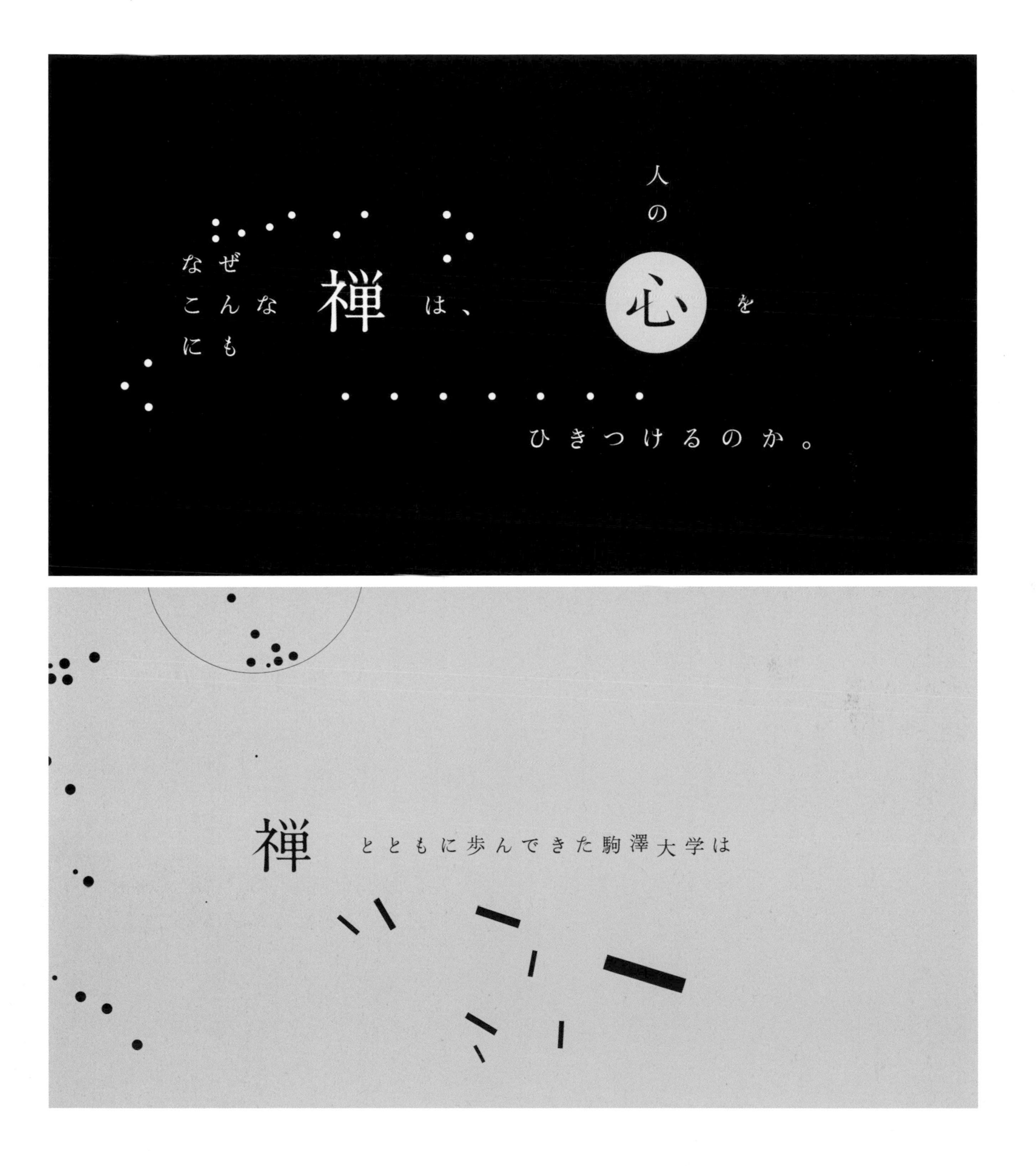
なぜこんなにも
禅は、
人の心を
ひきつけるのか。
禅とともに歩んできた駒澤大学は

Balticbest 2018

BY BOND

Balticbest is a boutique creativity festival, but now open for entries beyond the Baltic states. With this change of scale came the need for renewing the identity in a universal way to show the big ideas of small countries. To keep the focus clear, the visual language was stripped from any decors. Considering that digital screens are a primary communication platform, BOND used the black-and-white palette and no-nonsense grotesque Messina Sans to form the basis of the identity and marketing campaign. The simple but powerful motions play with the industry clichés and rely on the power of repetition, but at the same time constantly generate new layers and twists.

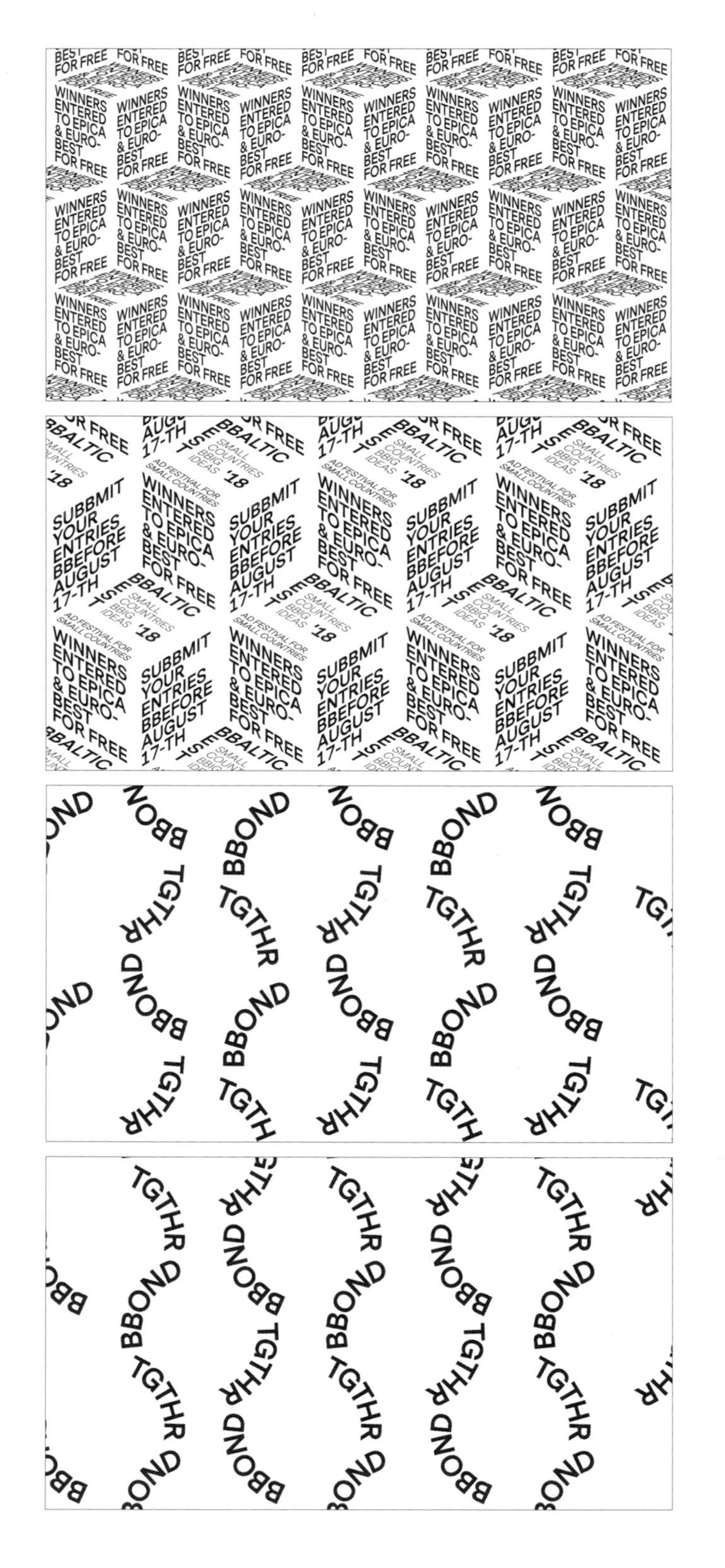

BALTIC BEST '18
BALTIC BEST '18
BALTIC BEST '18
BALTIC BEST '18
AUGUST 31, 2018
TALLINN, ESTONIA
BBALTIC

LO & BBEHOLD
BALTIC BEST '18
BALTIC BEST '18
BALTIC BEST '18
AUGUST 31, 2018
TALLINN, ESTONIA
BBALTIC

BB WELCOMED
BB WELCOMED
BB WELCOMED
AUGUST 31, 2018
TALLINN, ESTONIA
BBALTIC

STRATEGY **Nils Kajander** / DESIGN **Ivan Khmelevsky** / COPYWRITING **Ivan Khmelevsky, Nils Kajander** / CLIENT **Best Marketing**

Behance

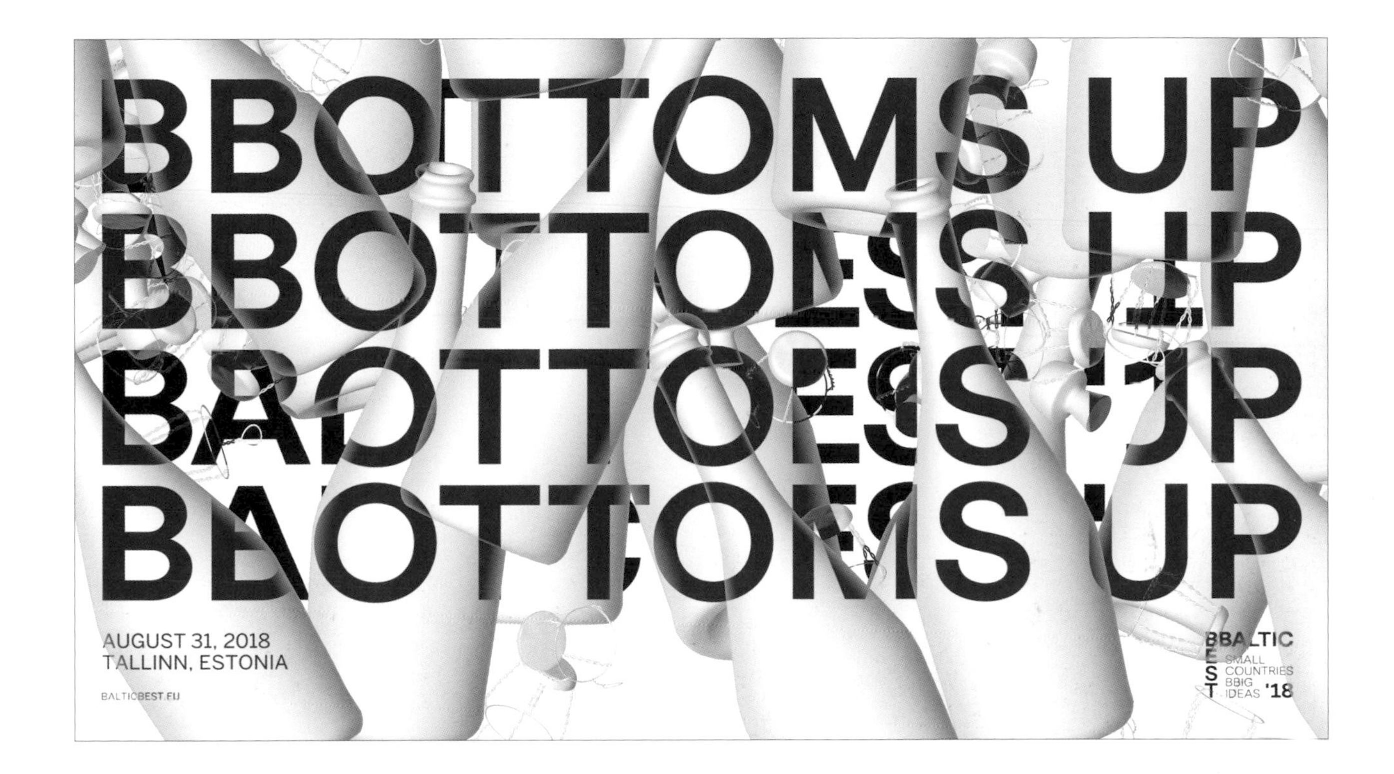

BALT BBEHOL18
BAL& BBEHOLD
BA& BBEHOLD
BO & BBEHOL8

AUGUST 31, 2018
TALLINN, ESTONIA
BALTICBEST.EU

BBALTIC
SMALL
COUNTRIES
BBIG
IDEAS '18

Blockchain

BY Fromsquare Studio

Extracting cryptocurrencies also known as mining requires the use of computer processing power, which leads consumption of huge amounts of energy. This animation designed by Fromsquare Studio presents ecological solutions for the blockchain technology and shows how renewable energy sources can provide new opportunities for the miners of cryptocurrencies. Meanwhile, the typography used in the animation goes perfectly with the geometric elements.

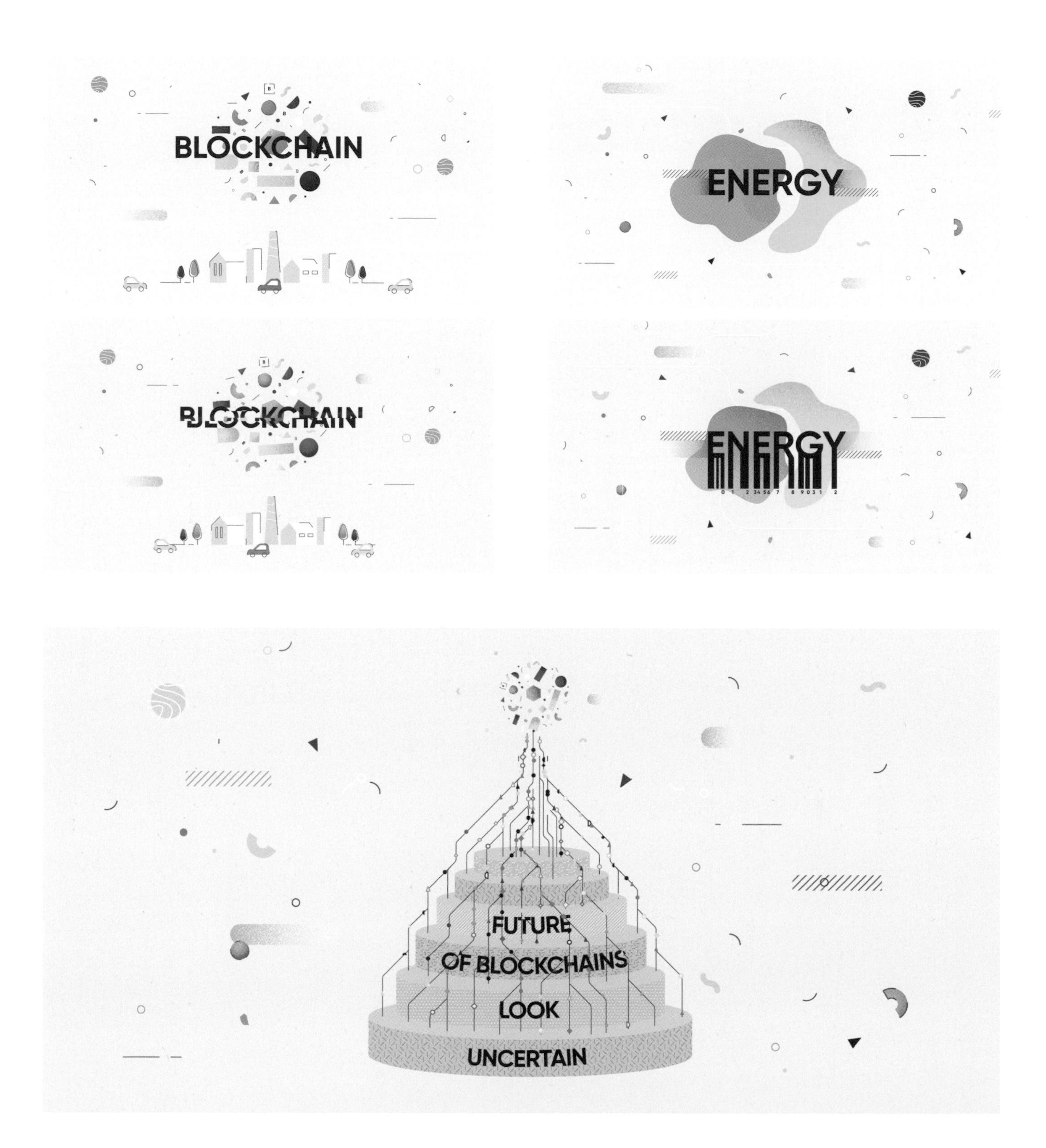

ART DIRECTION **Alicja Piotrowska, Kuba Piechota** / ILLUSTRATION & MOTION DESIGN **Alicja Piotrowska** / SOUND **Robert Ostiak**

Behance
Official

ICO
Initial
Coin
Offering

PROJECT Bi

BY MEMOMA Estudio

PROJECT Bi is an exchange and collaboration platform that offers incentives for the production and dissemination of national art and culture initiatives. At first, MEMOMA Estudio was asked to bring up a concept evocating Kandinsky's works and highlight the dot element in the entire project. With the minimalistic perspective, MEMOMA Estudio tried to reach the aesthetic and communication goals by creating style frames and sequences in a state of continual change. Finally, the entire video offers a playful and lively effect.

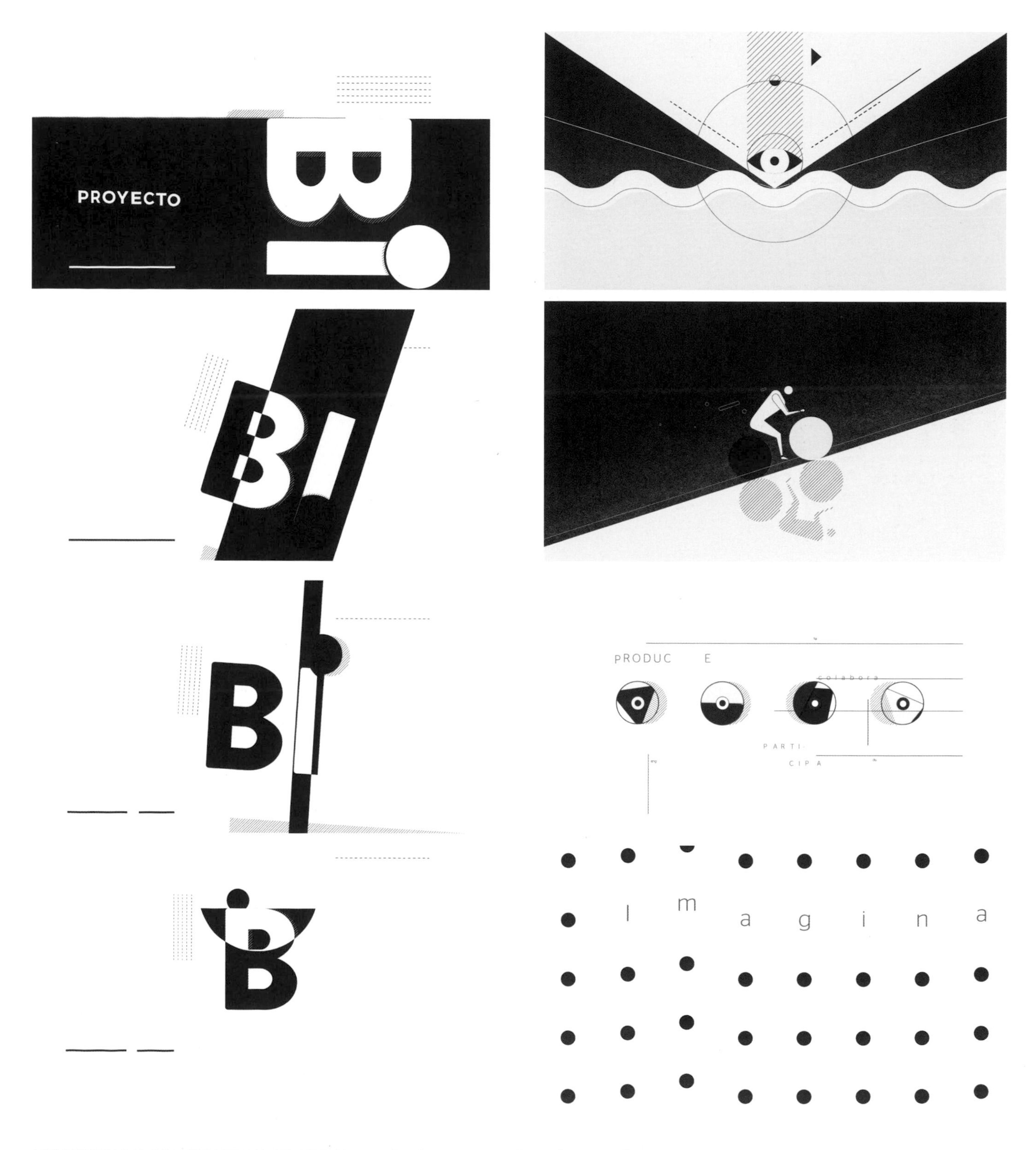

ART DIRECTION & ILLUSTRATION **MEMOMA Estudio** / CONCEPT **Alejandro Magallanes, MEMOMA Estudio**
MUSIC & AUDIO **Audio Magno**

Behance
Official

Imagina
COMPARTE
SÜ
+
mate

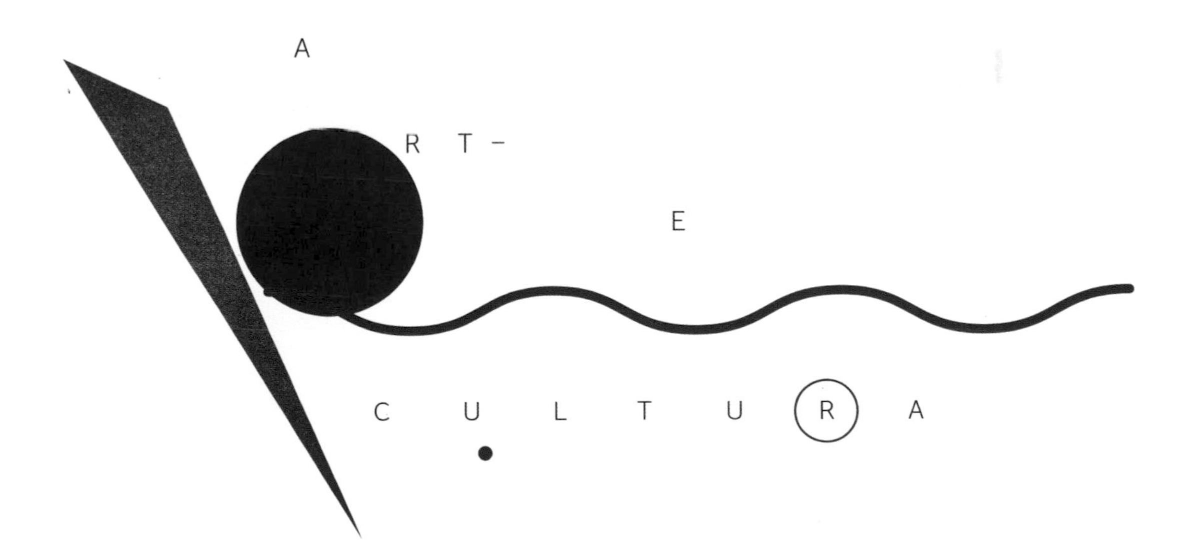
A
R T -
E
C U L T U R A

FIFA World Cup 2018

BY Xabi Mendibe

This video is a personal project for fun and playing with typography. According to the characteristics of different football terms, such as kick-off, yellow card, red card, and offside, Xabi made a series of GIFs initially. Also, these GIFs with short loops related to the events that took place during the matches in the FIFA World Cup 2018 in Russia and were shared in real time on social media with hashtags. When the World Cup came to an end, Xabi decided to integrate all the GIFs as a consecutive motion with sound design and some new scenes.

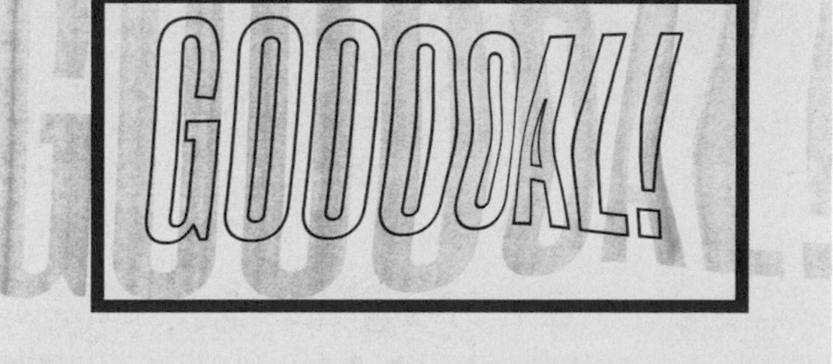

SOUND **Byron Abadia**

Behance

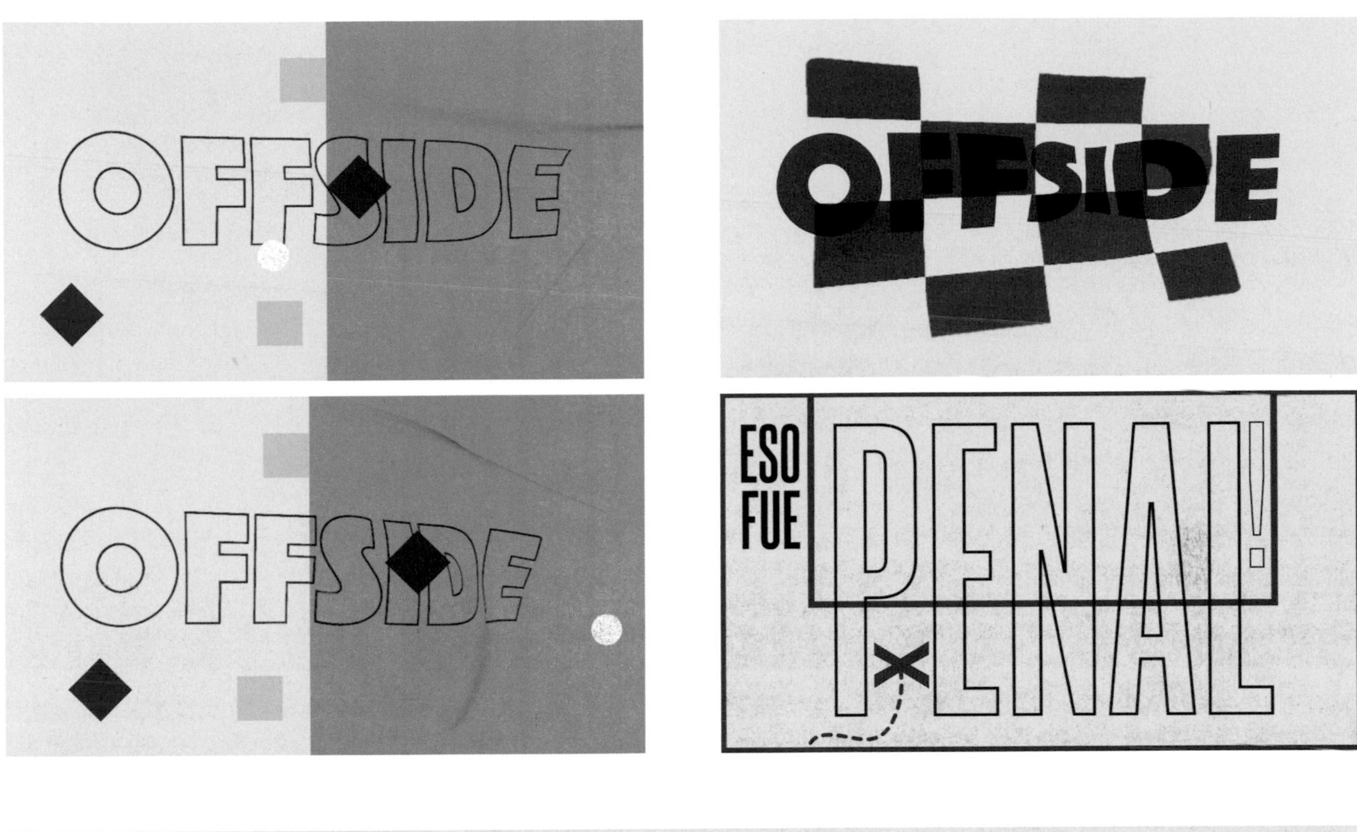
OFFSIDE
OFFSIDE
OFFSIDE
ESO
FUE
PENAL!
PENAL

THANK
FOR
FOR
FOR

FOX8 Rebranding

BY Lumiko

Lumiko was asked to rebrand FOX8. The FOX8's rebranding features bold and kinetic typography, iconic imagery, and playful style that can draw the attention of the viewers.

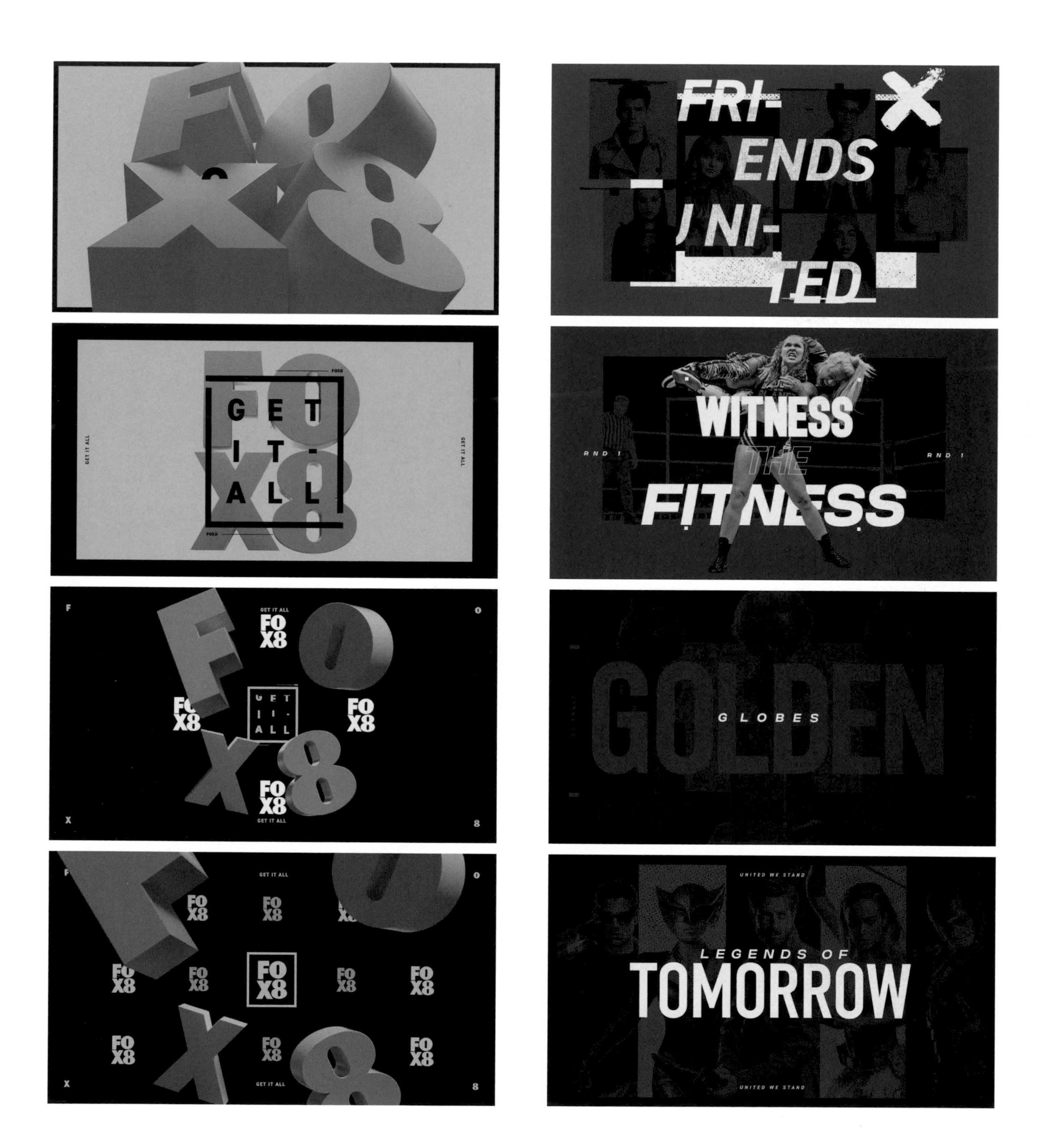

ART DIRECTION & ANIMATION **Chris Thompson (Lumiko)** / CREATIVE DIRECTION **Sean Vandeberg (FOX8)**
PRODUCTION **Mark Blondel (Where There's Smoke)** / MUSIC **Like Sugar (Chaka Khan)**

Vimeo

FOX8
FO
X8
FOX8

BEAM U HOTTIES
CHICAGO
FO
X8

SIMPSONS
FO
X8
THE SIMPSONS
TUESDAY 8.30
FOX8

Motion Motion Festival

BY NŌBL

With their multidisciplinary line-up, the Motion Motion Festival offers a panoramic overview of digital cultures and innovative creation. NŌBL was commissioned to create the identity of the 2018 edition. With structuring, deconstructing, and distorting the typography, this video was finally designed as a typographic moving composition that viewers can associate the manual work of typography.

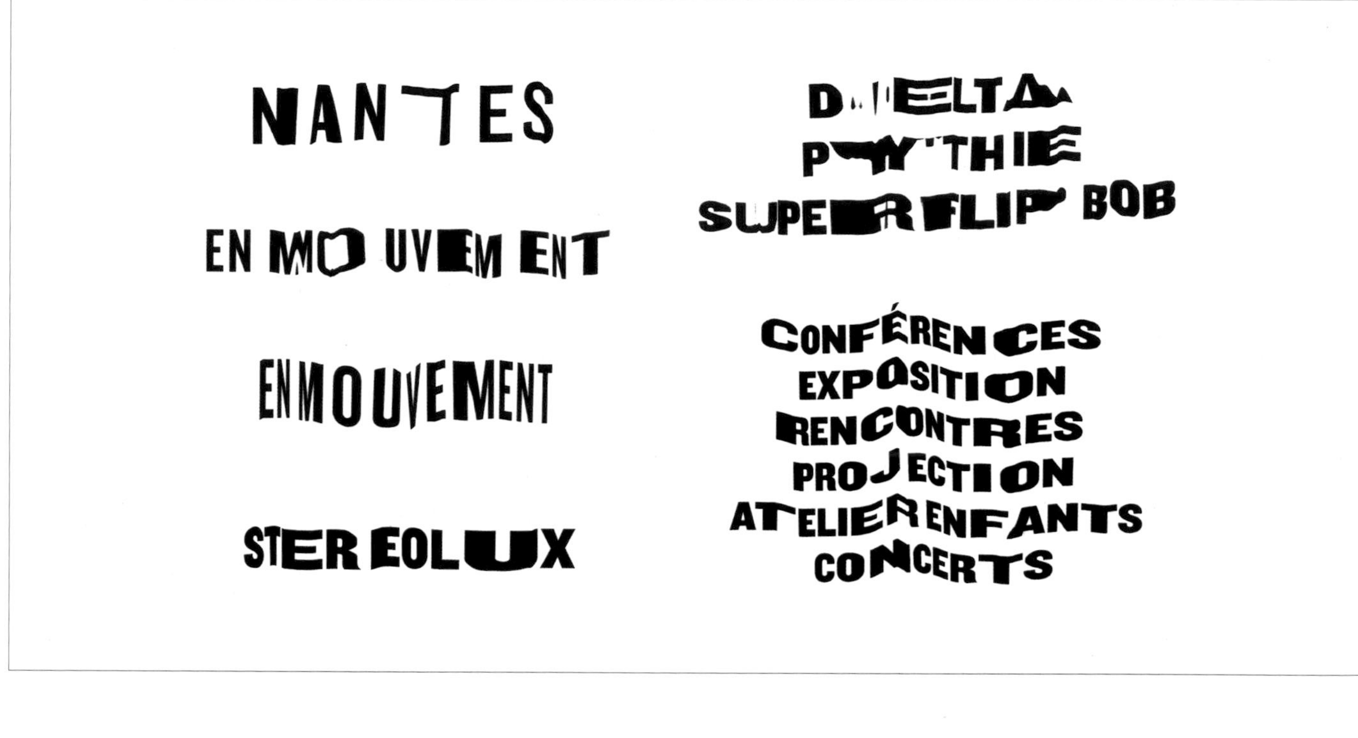

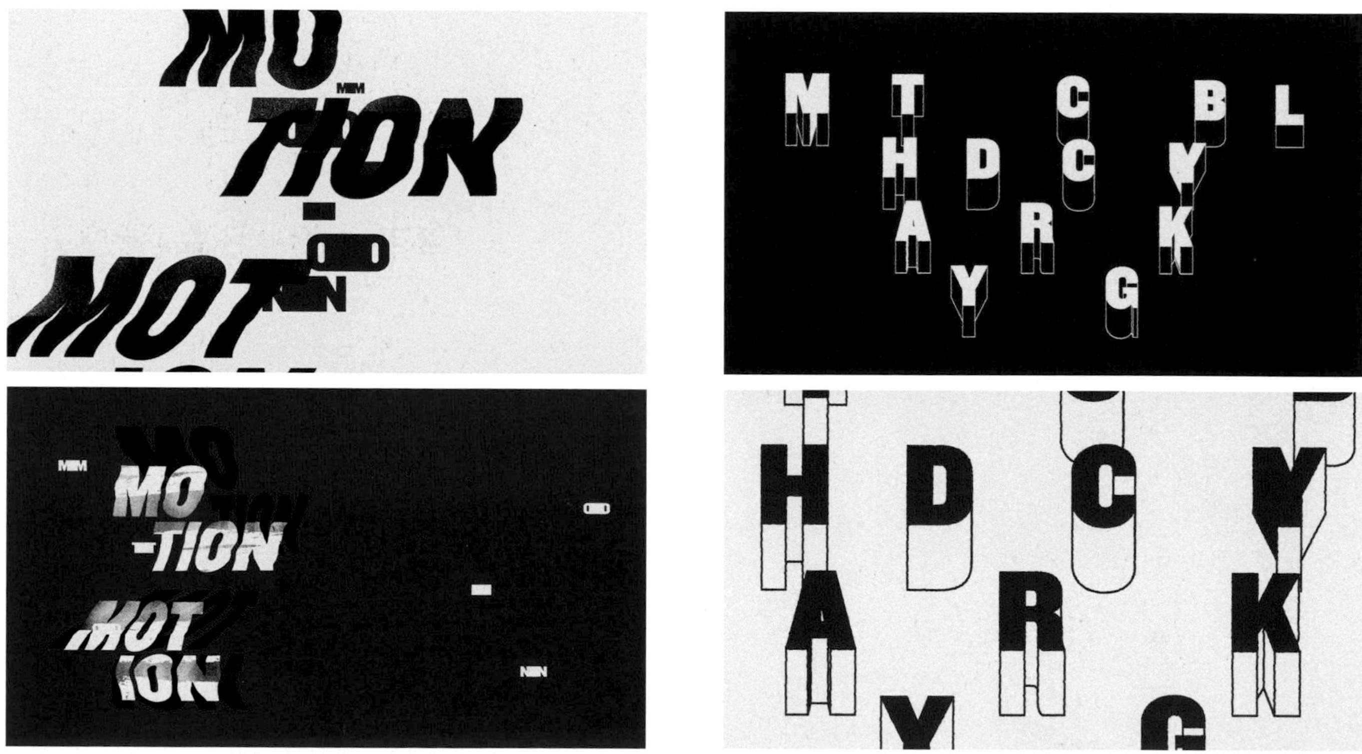

MUSIC & SOUND **Mooders** / CLIENT **Motion Motion Festival**

Official

O BINGO

LTA

I N L

N R O B N

PLAYGROUND

EXPOSITION

DELTA
OLIVIER RATSI
PYTHIE
PALEFROI & GUILLAUME MARMIN
SUPERFLIP BOB
BOB55

EXPOSITION

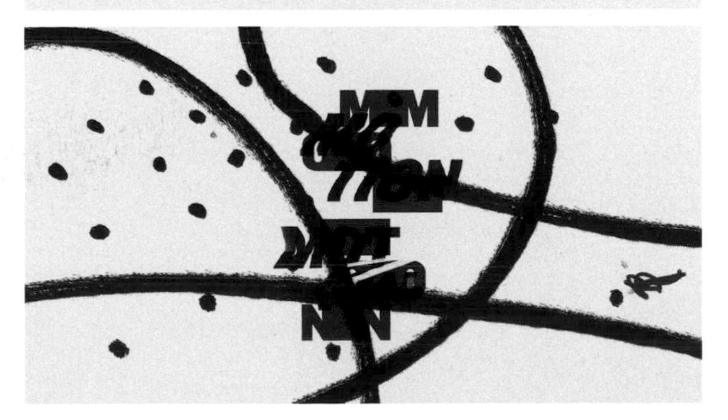

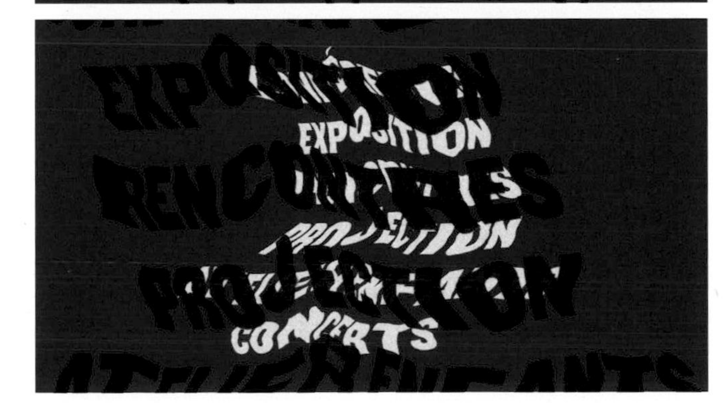

SURFACE

BY Frame

To coincide with the upcoming global tour of the SURFACE exhibition, Frame was excited to partner with the high-profile Danish photographer Søren Solkær. SURFACE is the culmination of a three-year journey whereby Søren captured over 150 of the world's leading figures in street art, including Shepard Fairey, Blek Le Rat, Seen, Ron English, Swoon, Faile, Space Invader, and many more. Employing a mixture of 2D and 3D design, Frame developed a new visual universe that seeks to capitalise the letters on the stunning array of images belonging to SURFACE, whilst adding a graphic twist that celebrates the certain aesthetics unique to street art.

DIRECTION, DESIGN & ANIMATION **Frame**

Behance
Official

(ENGLAND)
D*FACE

FRANCE
INVADER SPACE INVADER SPACE
BÄST
BÄST

Bigdrop

BY Javier Miranda Nieto

As an internal project in Bigdrop, this video was designed by Javier Miranda Nieto to play with the name "Bigdrop." The design was based on space and cosmos which are Javier's favourite. Getting inspired by the animation style of Ash Thorp and Jr.canest, Javier wanted to do something more structured in the typography. Meanwhile, his main idea was to use lines, guides, rules, basic figures, depth, and textures in the motion. The video was made all in Adobe Illustrator. Javier then used Photoshop to add all the textures and adjust the colour balance. When the still frames were ready, Javier took all the stuff to Adobe After Effects to animate. Then he used Adobe Premiere to add sound effects, music, and voice.

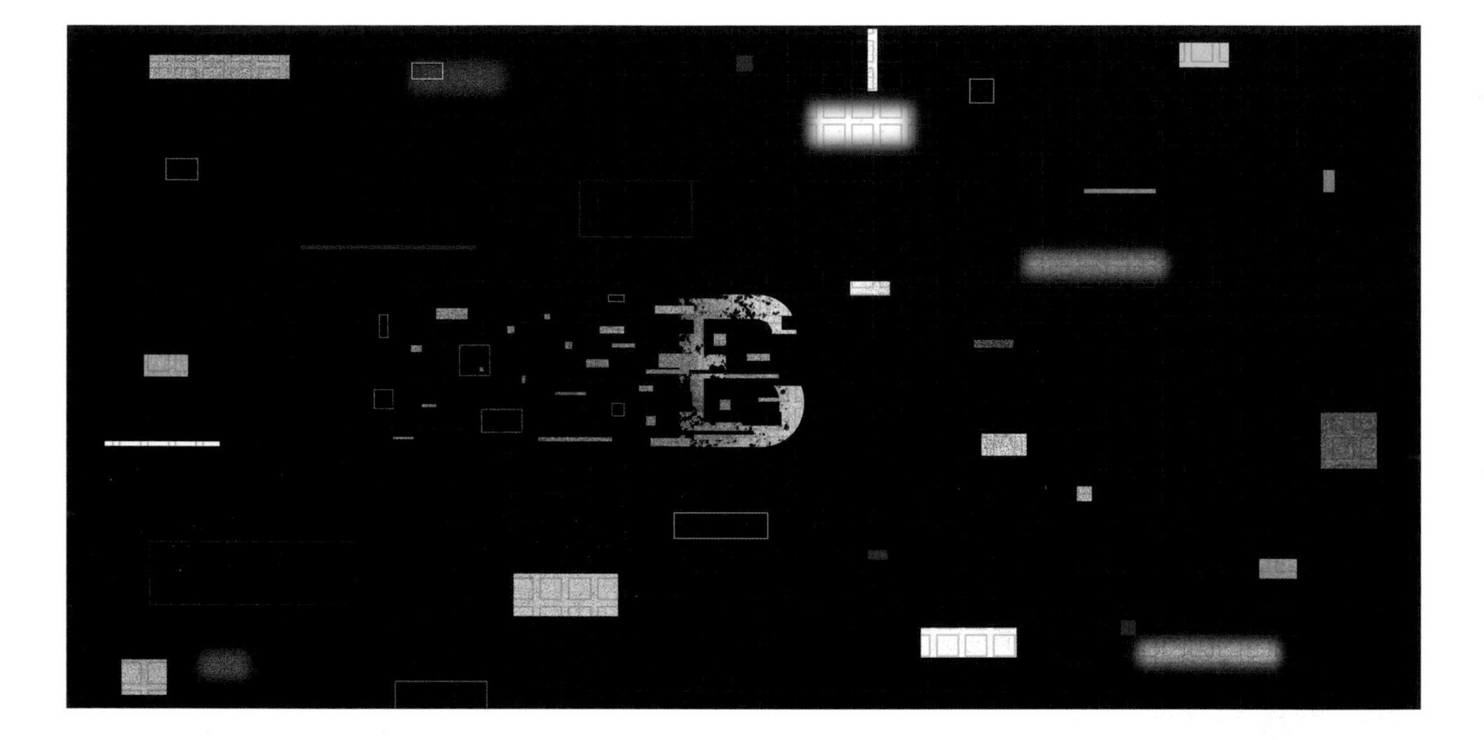

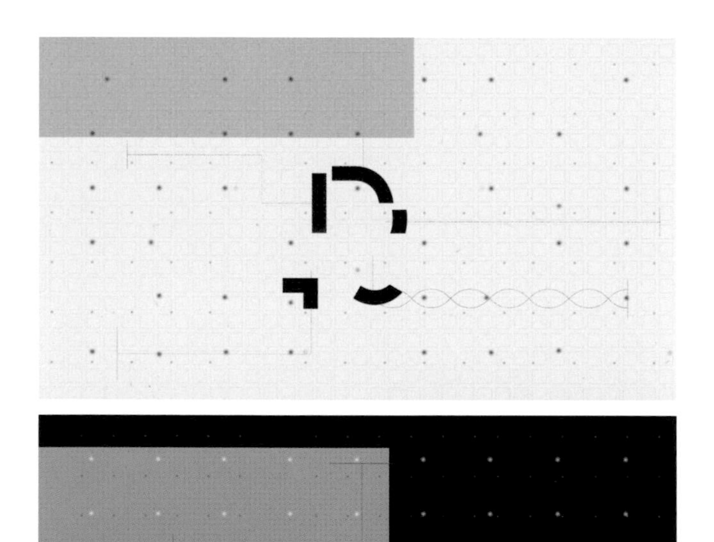

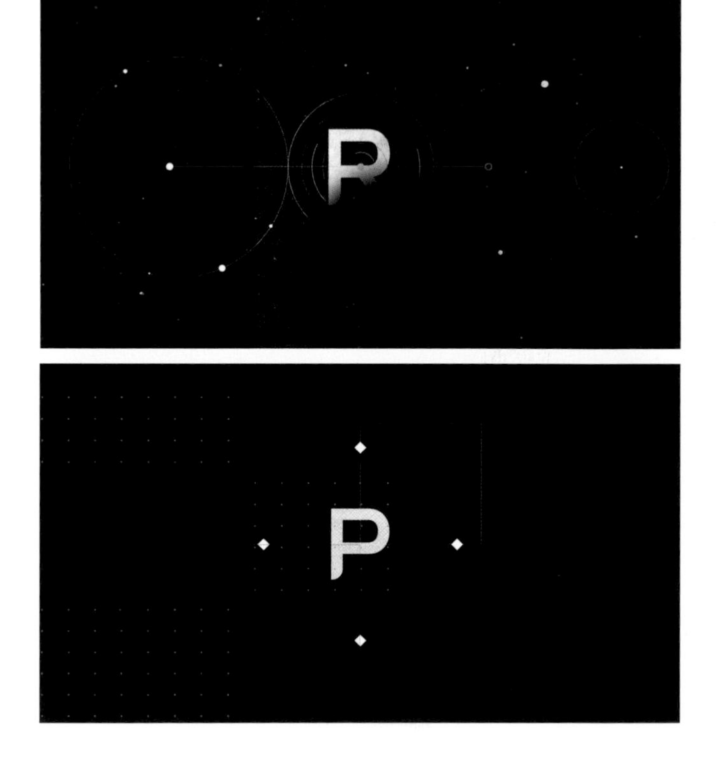

CLIENT **Bigdrop**

Behance

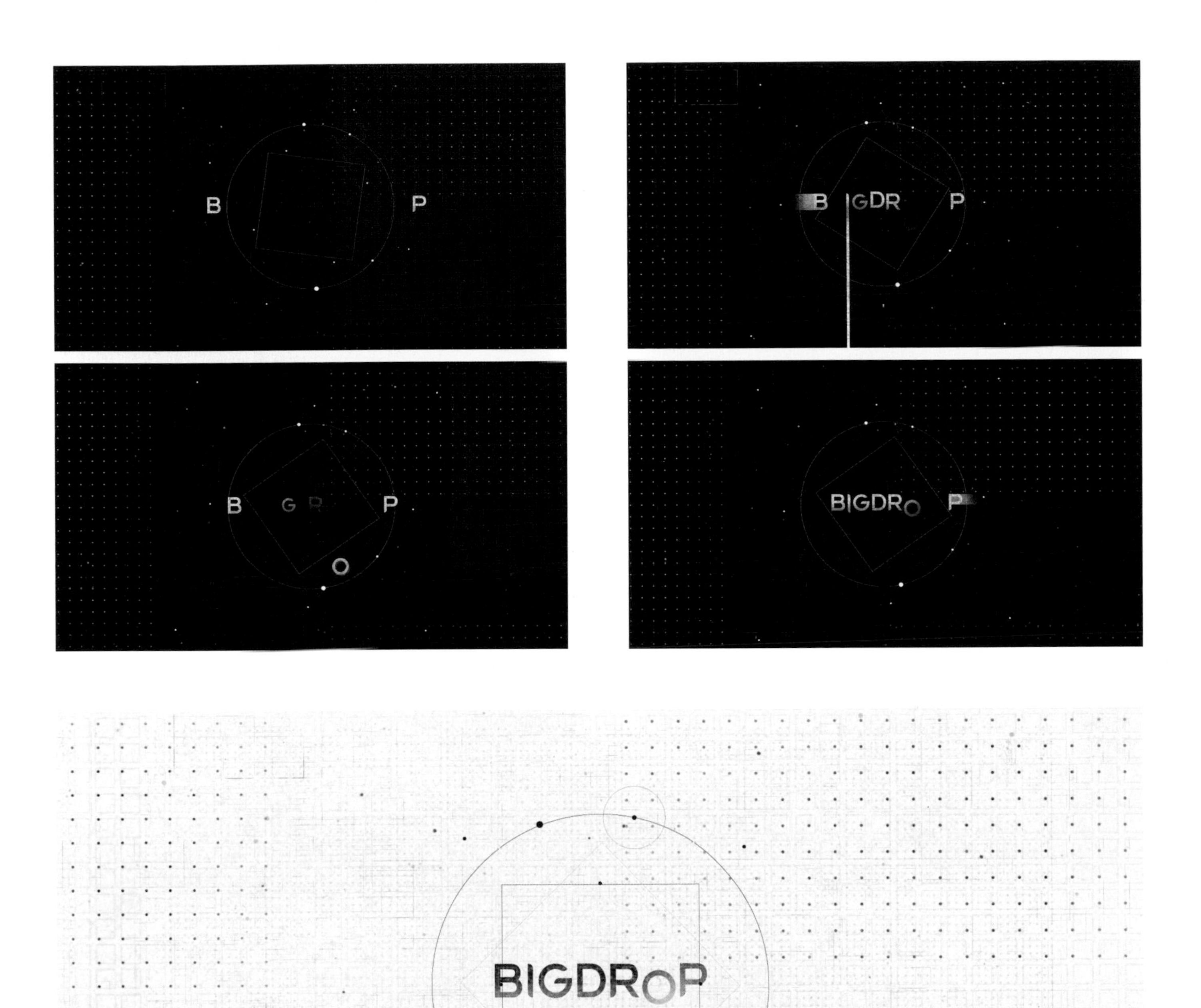
B
P
B
GDR
P
B
G R
P
O
BIGDRO
P
BIGDROP

Carbone Moving Identity

BY NŌBL

Carbone is a transmedia magazine dedicated to pop culture, creative literature, illustration, and interaction. Basing on the visual identity created by Superscript², NŌBL designed a movement-infused graphic charter for use with Carbone's video content on every platform. NŌBL got inspiration from the circular feature of the logo and the chemical element of carbon. Meanwhile, NŌBL designed the video's title by Stanley Font and subtitle by Styrene Font. Such two fonts have become the key elements in the motion principles that assist the blend of two contrasting components—the logo and the carbon element in a totally unique identity.

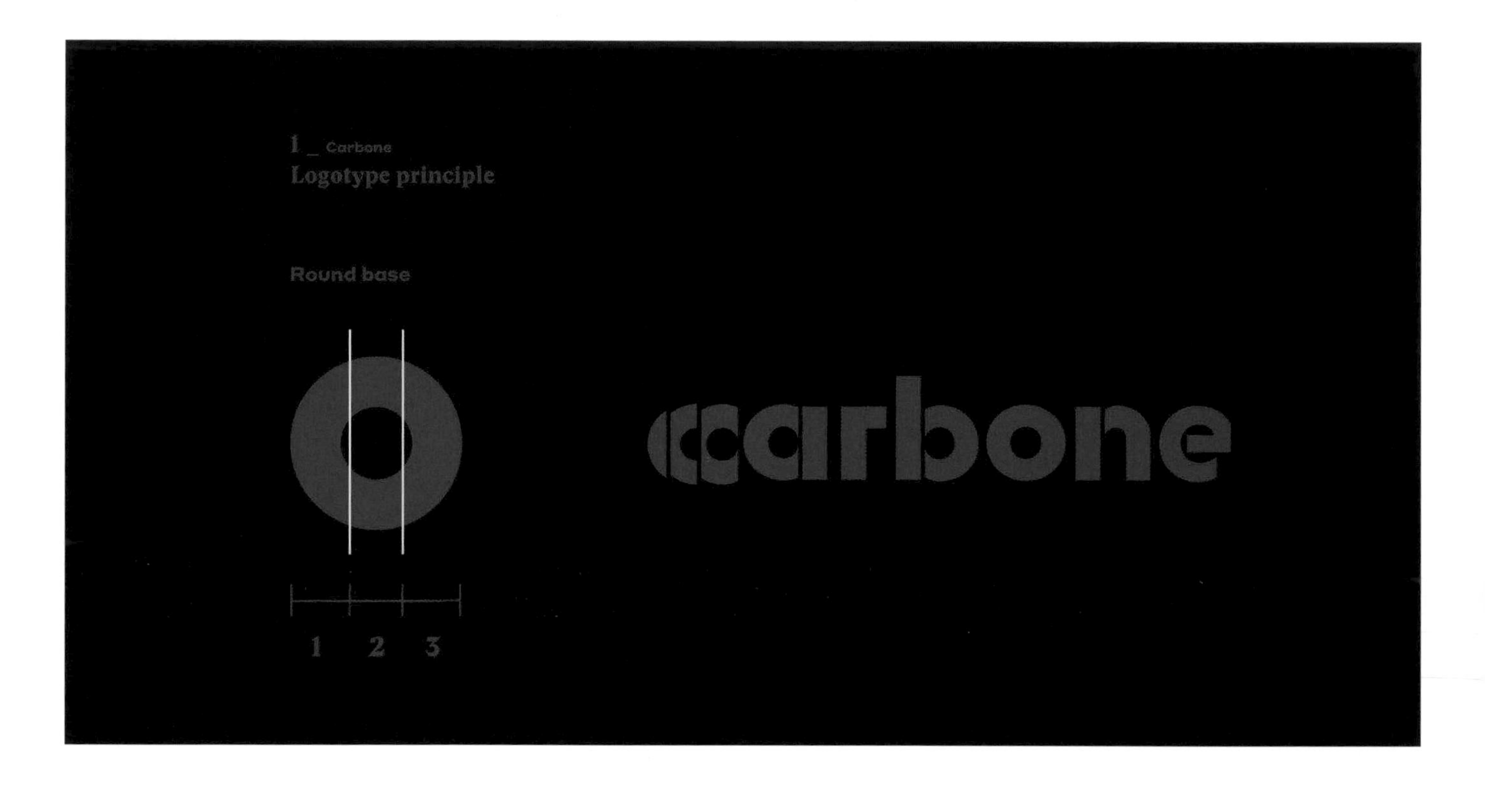

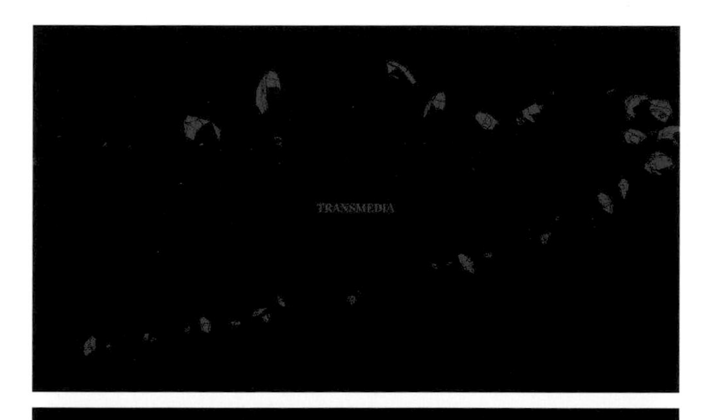

MUSIC & SOUND **Combustion** / CLIENT **Carbone**

Official

inspired by
PHILIP
K. DICK
Radioactive
DESERT
MEGACITY
overpopulated

FOMO

BY Kurppa Hosk

FOMO is a web-TV live broadcast, a concept, and the brainchild of Pontus Djanaieff and Jonas Kleerup. When FOMO approached Kurppa Hosk with their idea, it was immediately evident that the brand identity and visual communication needed to be as visually stimulating as the content. A reoccurring theme during the collaboration was giving visitors of the platform the notion that they were being absorbed entirely by the content. Moreover, the new design would have to tirelessly produce the feeling that there is a constant supply of energy.

Fomo
Increasing Frequency (v)

10^{02} X–rays

10^{04} SLF
Super–low frequency (radio)

VF
Voice frequency

10^{06} LF
Low frequency
(radio)

10^{08} HF
High frequency
(radio)

10^{10} UHF
Ultrahigh frequency
(radio waves)

10^{14} EHF
Extremely
high frequency

FOMO TV
We have the pleasure of
inviting you to the
FOMO launch party
Thursday 01.21, 21.00–
RSVP at
+46-76-174-63-56
Email pontus|at|fomo.se
www.fomo.se
www.youtube.com/fomo

CLIENT **FOMO**

Behance
Official

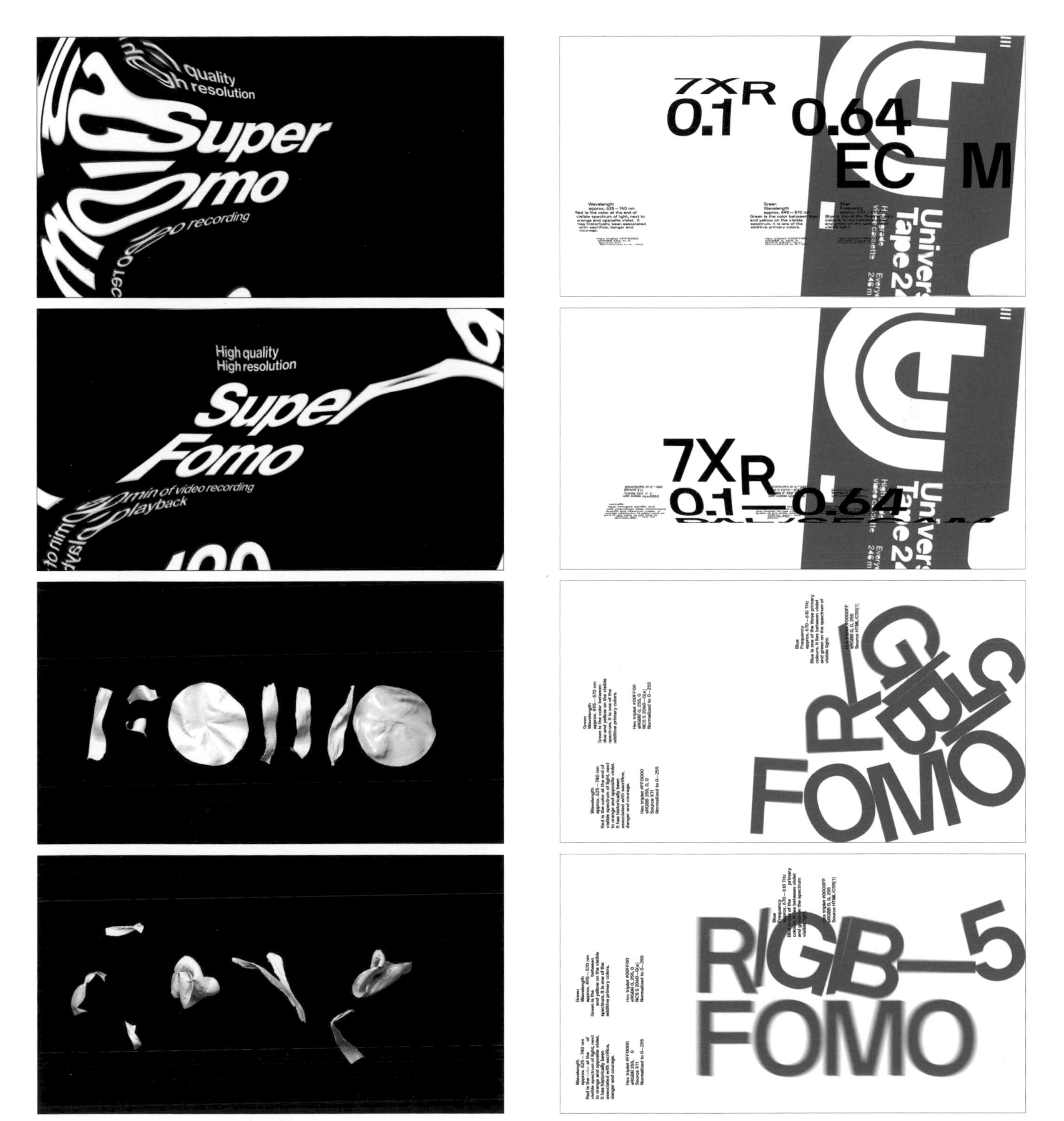
quality
h resolution
Super
mo
High quality
High resolution
Super
Fomo
min of video recording
Playback
7XR
0.1 R 0.64
EC M
Univers
Tape 24
RGB—5
FOMO

Discovery Truth Files

BY A1 Design

Discovery Truth Files reveals history for the smart, passionate, and deeply curious audience. A1 Design has developed a new visual identity for this brand. It goes deeper into stories with a pursuit for innovation. A1 Design chose Compacta that speaks bold and loud in the video. When the viewers see the simple and clear headlines, they can easily trace the overall graphics style from newspapers to today screens.

ANIMATION **A1 Design** / CLIENT **Discovery Networks**

Vimeo

BOB
WOODWARD
&
CAR
BER
Ford praises Nixon speech
NIXON
DESIGNS
FBI
SPECIAL AGENT
LOADING
62
PREMIO
PULITZER

FINAL
The Washington Post
15¢
New York, N.Y. 10017, Friday, August 9, 1974
NIXON
RESIGNS

Alternative Movie Title for Contact

BY Sonya Nechkina

Contact is an American fiction drama film directed by Robert Zemeckis in 1997. And Sonya Nechkina designed this movie title as a training exercise. The main task was to use the typography as the main mean of expression. As a result, she decided to create an alphabet of icons that turn into letters and then create titles. Each icon firstly transforms into a diagonal line from which the letter appears.

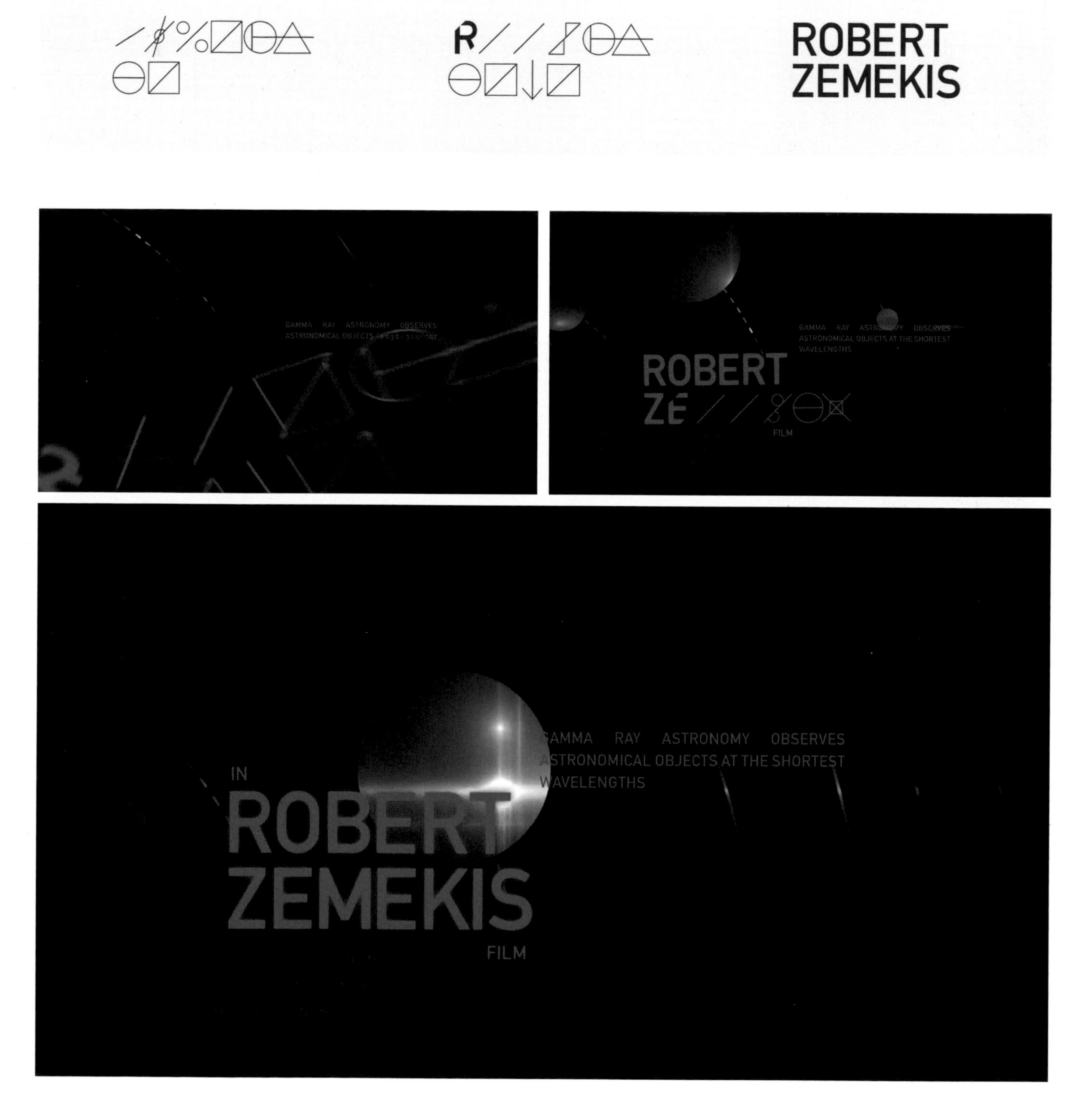

CONCEPT & ANIMATION **Sonya Nechkina**

Behance
Vimeo

MATTHEW
MCCONAUGHEY
AS PALMER JOSS
CONTACT

DNC 2018 Theatre Kikker Trailer

BY True Form

Theatre Kikker in Utrecht, Netherlands asked True Form to design a trailer for their new festival called De Nieuwe Collectie. The theatre's typographic posters made by Herman van Bostelen were the visual basis for this trailer. True Form transformed the unique rhythm of freshly graduated theatre artists into kinetic typography and showed this distinctive festival with a feeling of old-school theatre-making.

CLIENT **Theatre Kikker**

Vimeo

SPANNEND

R S T E L L I N

DE SPANNENDSTE
VOORSTELLINGEN

VAN DE NIEUWSTE
THEATERMAKERS

Neu Forma Grotesk Font Presentation

BY Alexander Efremov

This video was made for Typomania 2019—an international typography festival—and pays a tribute to the intro of the *Cowboy Bebop*—a Japanese sci-fi animation.

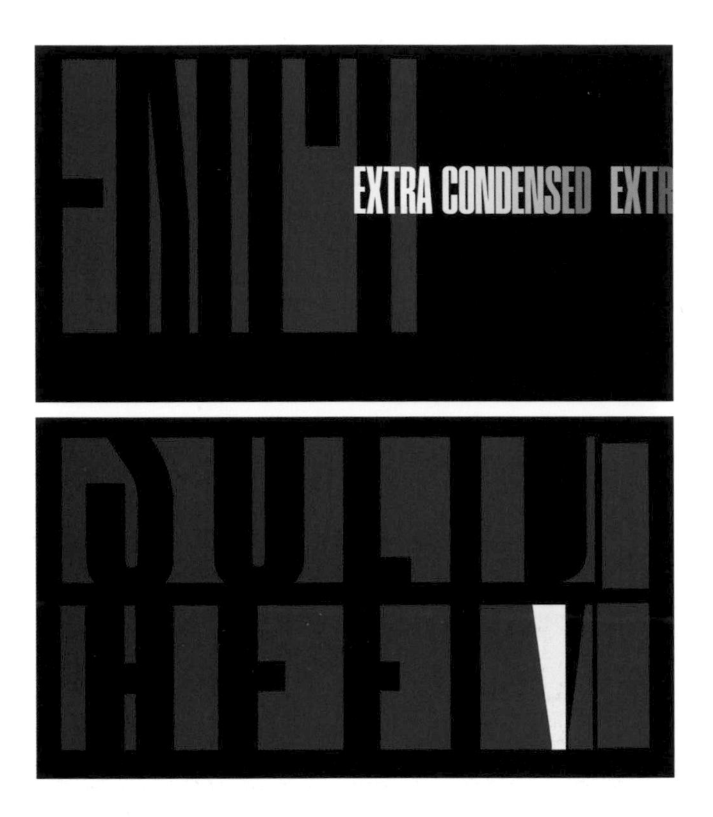

MOTION DESIGN **Alexander Efremov** / TYPOGRAPHY DESIGN **Timur Zima**

Vimeo

À Á Â Ã Ä Å Æ

Run.wav Opening Title

BY Studio BYTS

This video was designed for the TV music show program "Run.wav." Studio BYTS combined kinetic typography with the elements of the actual stage's set design of the music show.

DESIGN & ANIMATION **Isaac Hong**

Vimeo

RUN.
RUN.
RUN.

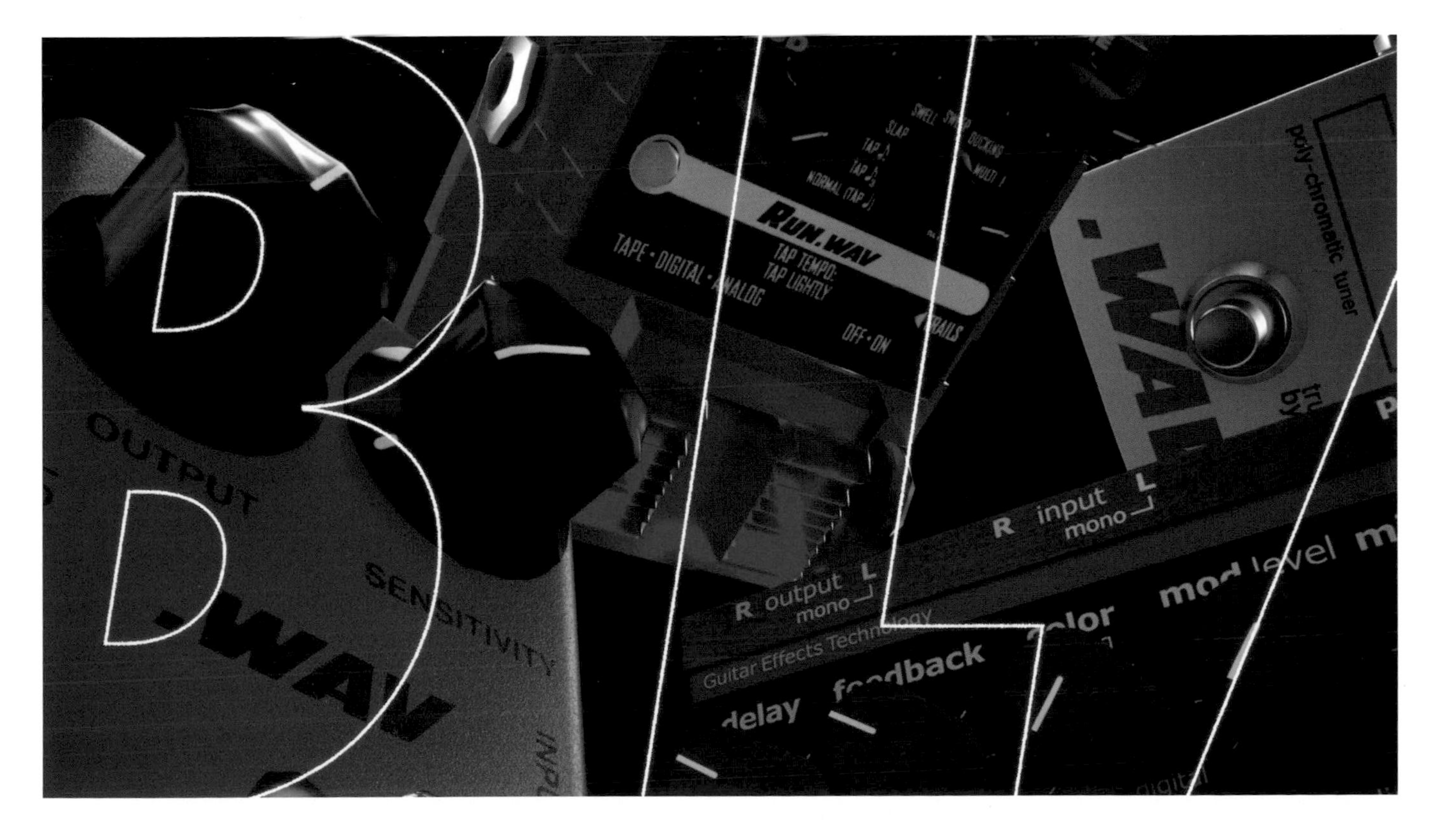

SWELL
SLAP
DUCKING
MULTI
RUN.WAV
TAP TEMPO:
TAP LIGHTLY
TAPE • DIGITAL • ANALOG
OFF • ON
TRAILS
poly-chromatic tuner
OUTPUT
SENSITIVITY
.WAV
R input L
mono
R output L
mono
Guitar Effects Technology
delay
level

Mutek 2019 Promotional Trailer

BY Sebastien Camden

This promotional trailer for the international electronic music festival Mutek puts a big emphasis on dynamic type animations over various 3D renders of abstract objects and textures. The designers' objective was to create a fast-paced and enticing video that would make the viewers want to watch it again as well as build hype at the same time.

AGENCY **Commissaire** / CREATIVE DIRECTION **Commissaire** / DESIGN **Zacharie Lavertu, Duc Tran, Sebastien Camden**
ANIMATION **Sebastien Camden, Aaron Kaufman, Duc Tran** / CLIENT **Mutek MTL**

Vimeo

Party Sleep Repeat 2019

BY Check it Out!

This promotional video is created for the music festival Party Sleep Repeat. The designers started working from the initial designs of 2019. They wanted to play with typography to explore different ways and techniques to animate the bands' names and sync them with Dead Combo's music *Deus me dê Grana*.

DESIGN **Gonçalo Antunes, João Pessegueiro, Carlos Marques** / ANIMATION **Carlos Marques, João Pessegueiro**
MUSIC **Dead Combo** / CLIENT **Associação Cultural Luís Lima**

Official

MELQUÍADES

NOTYOURSERIES 2
Teaser Video

BY Studio BYTS

NOTYOURSERIES is a cultural platform, aiming to introduce unknown cultural contents through various events, such as concerts, parties, and exhibitions, and provide a cultural experience. In this promotional video, the title NOTYOURSERIES became a key visual expressing the characteristics of this platform with a bit of humour.

Vimeo

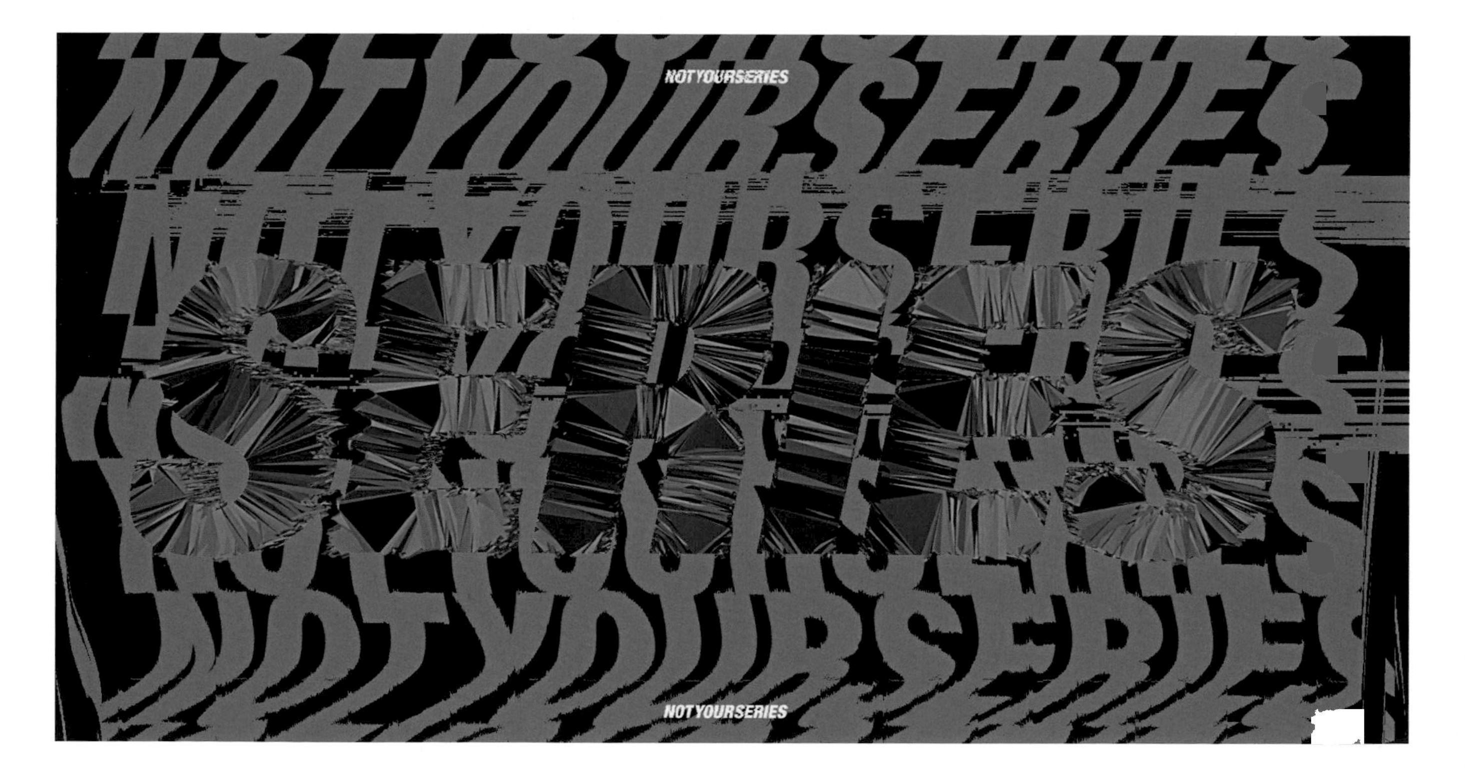

DESIGN & ANIMATION **Isaac Hong**

NOTYOURSERIES 3 Teaser Video

BY Studio BYTS

In this video Studio BYTS used the main colours from NOTYOURSERIES 3 and created the kinetic typography with the effects of glitch and displacement.

Instagram

DESIGN & ANIMATION **Isaac Hong**

MTV MIAW 18

BY Pes Motion Studio

This brand identity of MTV MIAW 18 was based primarily on the typographic acronyms widely used in social media, such as LIT, FOMO, and AF. And the chosen typographic treatment was the result of experiments that combined spatial movement with repetition and rhythm.

AGENCY **Pes Motion Studio** / CREATIVE DIRECTION **Pedro Estanga** / PRODUCTION **Paula Chapela**
CLIENT **MTV Latin America** / CREATIVE DIRECTION **Sean Saylor, Edson Fukuda** / MANAGEMENT **Alexandra Congrains Pender**
PRODUCTION **Germán Verdi** / ART DIRECTION **Germán Verdi, Vinicius Prado** / DESIGN **Roberto Ramírez** / AUDIO **Ricardo Ramirez**

Official

L
I
K
E
LIT
LIT
LIT

SORRY NOT SORRY

Webby Awards Identity

BY Bekk, Babusjka

The Webby Awards is a renowned international design competition that each year brings together the very best of the design industry from all over the world to celebrate works of excellence. The 2019 year's concept plays upon "Internet we want." As the Internet is dualistic and a mosaic, the goal with the new identity and production was to visualize both the positive and negative aspects, imply that they are the two sides of the same coin. The award show's opener would kick off the whole show and set the mood for the rest of the night. This opener takes the audience on a kinetic typographic journey through the Internet, using Internet terms that describe the good and bad sides of it.

Webby
Awards

CREATIVE DIRECTION **Hans Christian Øren, Bekk** / VISUAL IDENTITY DESIGN **Bekk** / MOTION DESIGN **Babusjka**
SOUND **Petter Haavik, Sagveien Resort** / CLIENT **Webby Awards, Jeff Zemetis**

Behance

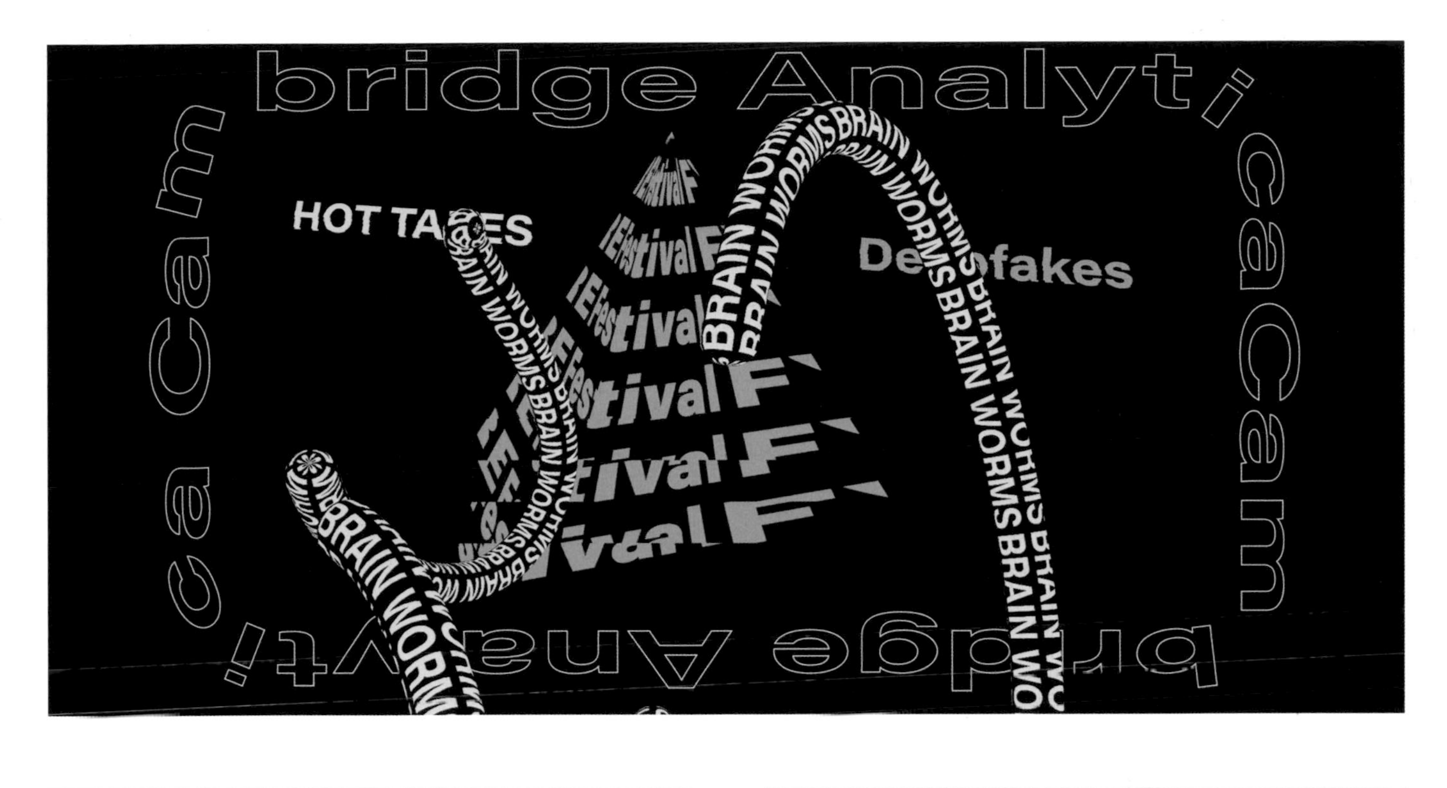

bridge Analyti
HOT TAPES
Deepfakes
BRAIN WORMS
Festival

STANSTANSTAN
CONTENT
NETFLIX
CHILL

Hohaiyan Rock Festival Image 2018

BY Whitelight Motion

As an annual rock music festival, the 18-year-old Hohaiyan Rock Festival was celebrating its own coming-of-age ceremony. The word "music" in Chinese is a heteronym which also means "happiness." This video interprets this heteronym in order to display the atmosphere of happiness in the music live of Hohaiyan Rock Festival. Combines with rough strokes and wave ripples, the design of typeface also reflects the open and inclusive characteristics of Fulong Beach where the music live was held in.

DIRECTION **Rex Hon** / STYLE FRAMES **Tim Tseng** / DESIGN **Kai Chun Hsu, SihPei Wu, Morris Chi**
MUSIC & SOUND **Hsu Chia-Wei** / MANAGEMENT **Sam Shen**

Official

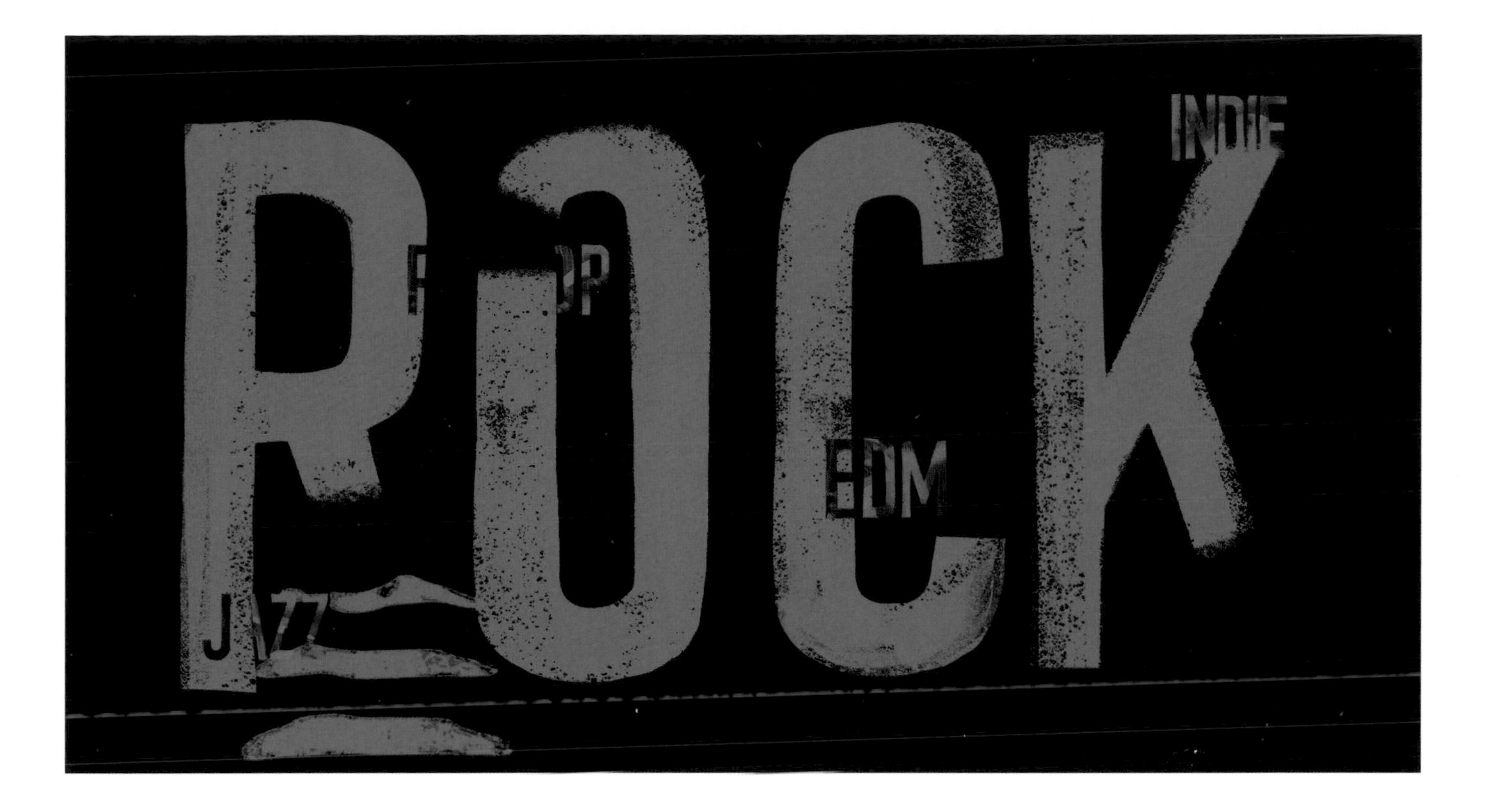
ROCK
INDIE
POP
EDM
JAZZ

JAZZ
PUNK
POP
INDIE
EDM
PUNK

MUSIC
MUSIC

Catfish Colombia S2 Promo

BY Pes Motion Studio

Pes Motion Studio once again worked with MTV Networks Latin America to develop the image campaign for another season of *Catfish Colombia*. For this project, the typography treatment is a combination of digital chat text boxes in social media and an effect generated by the glitches and drops that happen when the handheld cameras expose catfishing (a type of deceptive activity where a person creates a fake social networking presence for nefarious purposes).

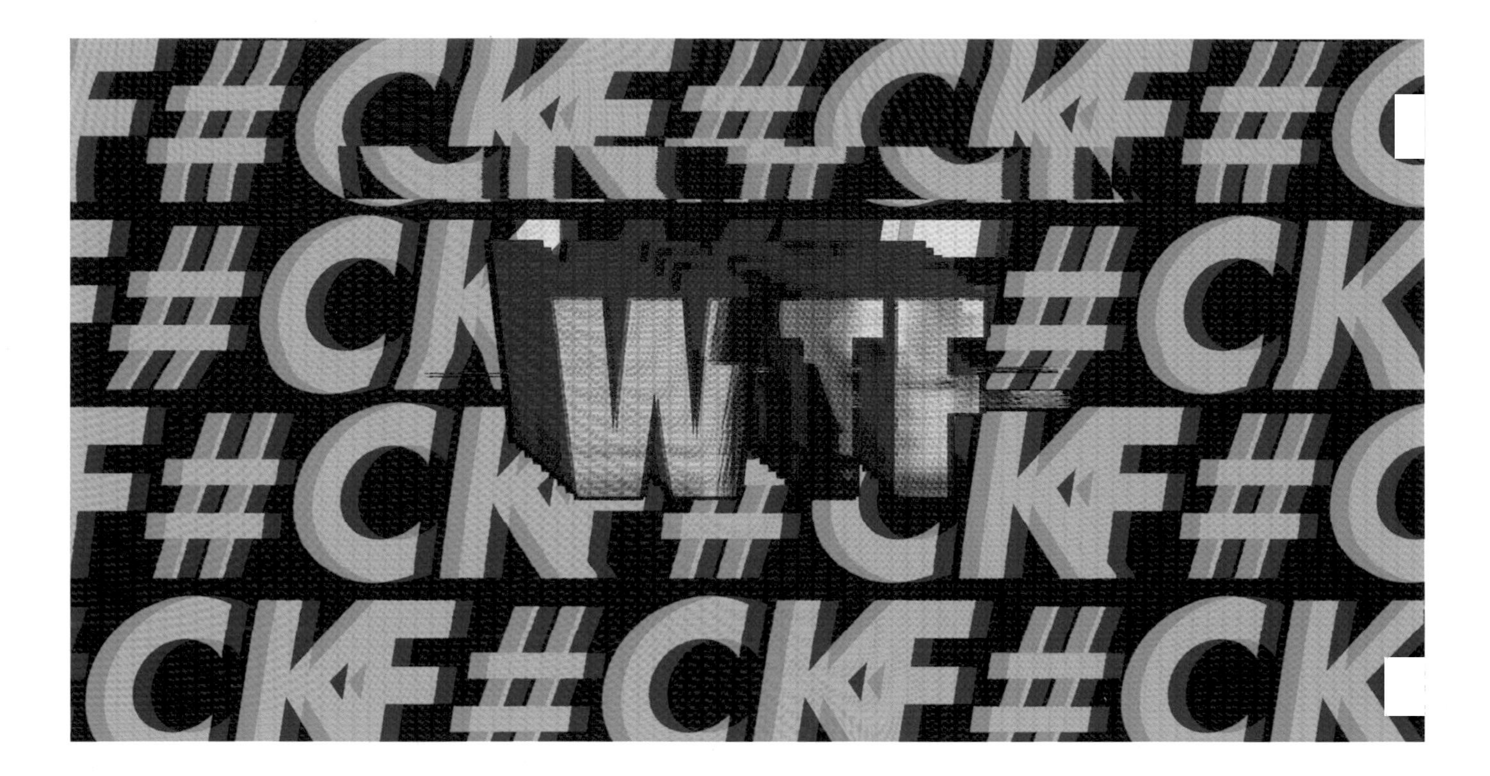

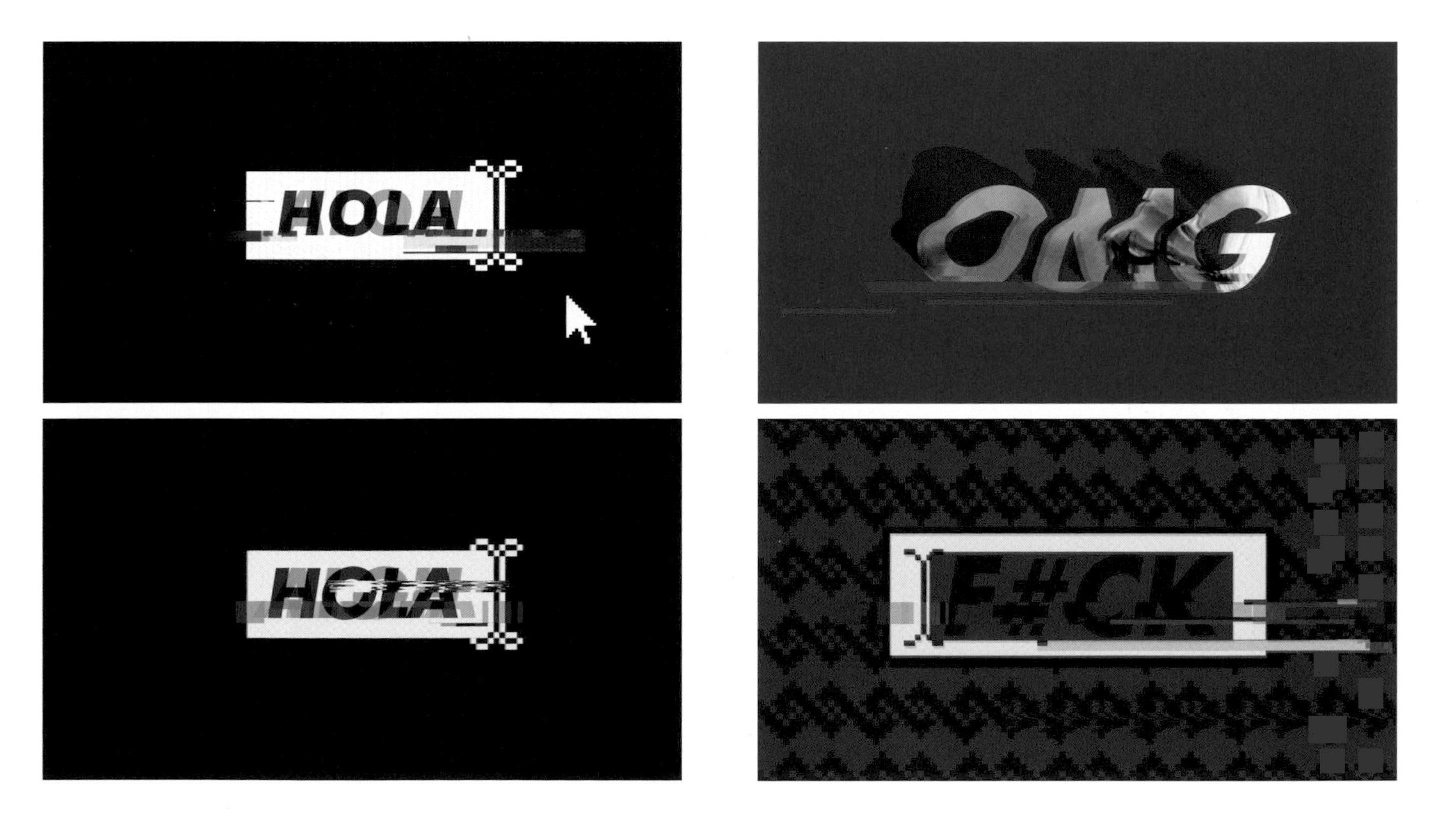

AGENCY **Pes Motion Studio** / CREATIVE DIRECTION **Pedro Estanga** / PRODUCTION **Paula Chapela**
CLIENT **MTV Latin America** / CREATIVE DIRECTION **Sean Saylor** / MANAGEMENT **Alexandra Congrains Pender**
PRODUCTION & ART DIRECTION **Germán Verdi** / COPYWRITING **Gloria Concepción**

Behance

BOMA

BY Papanapa

BOMA gets connections with its audience through an experience driven by music. Papanapa had to reflect the authenticity and surprise of such connections so that BOMA can establish a true and desirable relationship with its audience. Papanapa specially designed a monospaced font for BOMA that evokes music and excitement. The characteristic of this font displays vibration and flexibility in motion graphics. The wordmark and identity were built by a layered system of graphic elements that can be applied in many different ways to create a vivid and exciting visual communication.

CLIENT **Entourage**

Behance
Official

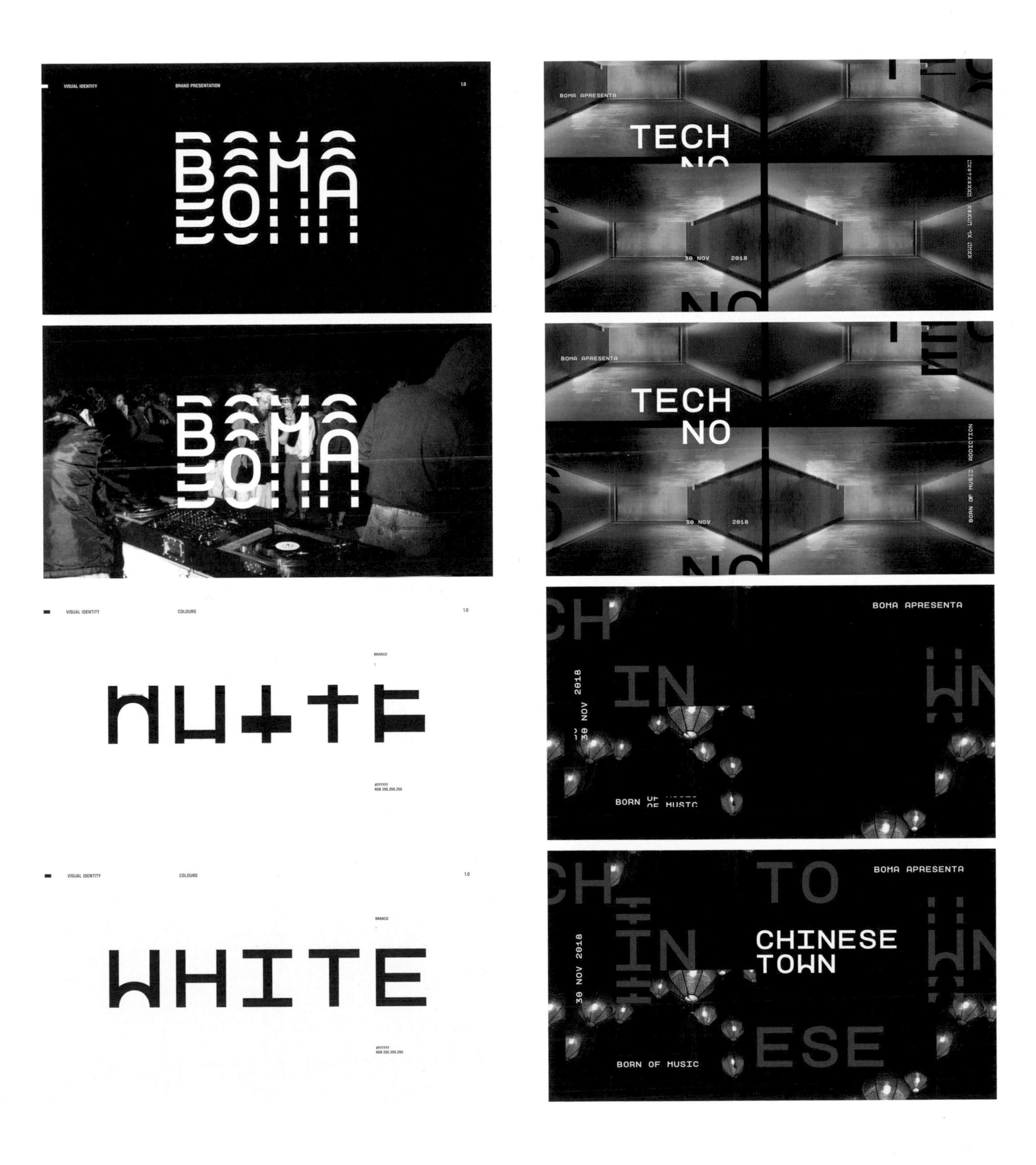
BOMA
BOMA APRESENTA
TECH
NO
30 NOV 2018
BORN OF MUSIC ADDICTION
WHITE
BOMA APRESENTA
30 NOV 2018
BORN OF MUSIC
CHINESE
TOWN

Summer, an Island

BY Manuel Martin

Summer is an island between two seas—spring and autumn. This project aims at turning the concept of summer into a motion graphic with a tense feeling of impending finality. Without the typical "fun in the sun," nostalgia and imminence are the main themes. The images of delights, excess, pain, smell, and adventures based on Manuel Martin's life in Venezuela. The overall aesthetic is the result of the initial experiments, such as the glitch effect and the stretch of the typeface.

Behance
Official

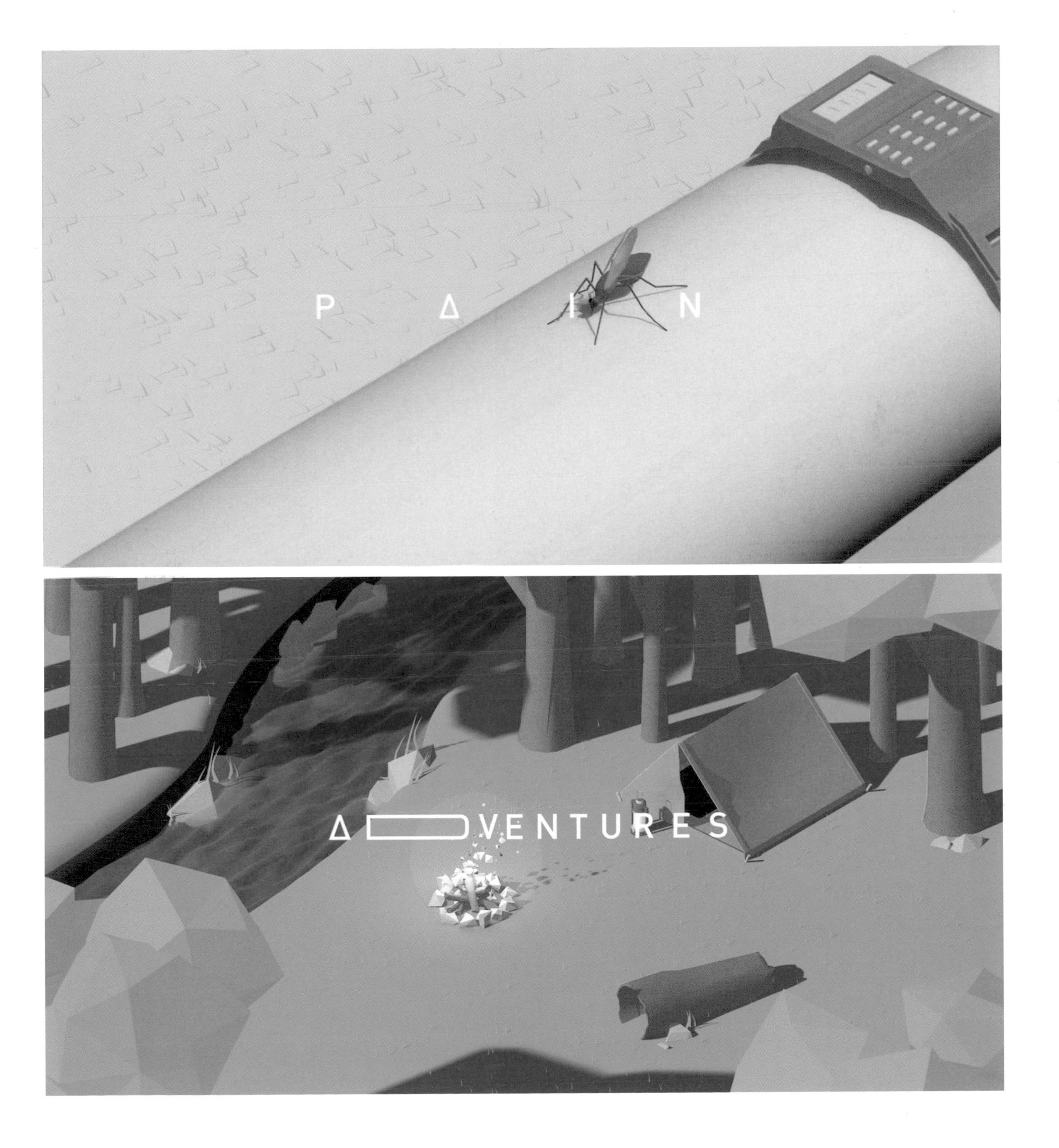
P A I N
A VENTURES

HNY_19

BY Vincent Dumond

This personal project based on the message "Happy New Year 2019." Vincent Dumond designed the animation of Renaud Futterer's variable font, evolving between 2D and 3D geometric shapes.

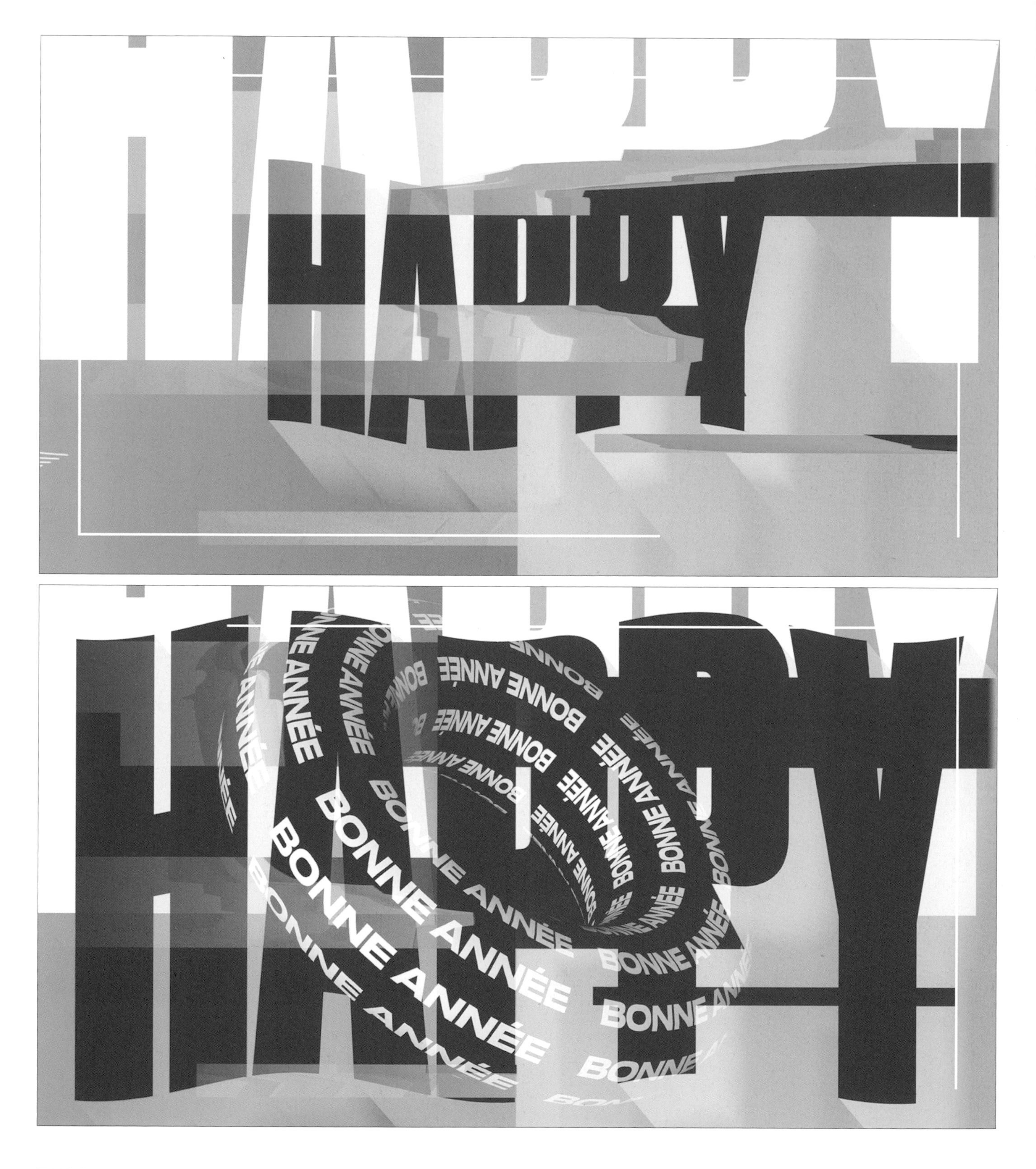

TYPOGRAPHY DESIGN **Renaud Futterer** / SOUND **Jean-Baptiste Saint-Pol**

Vimeo

Sigfox "0G World" Open Title

BY Vincent Dumond

This jingle animation was created for Sigfox—a start-up company working on the 0G concept. The animation, relating to keywords, visually translates the interaction and frequency of 0G technology thanks to fractals, waves, and rounded shapes. The background was built on the colour chart range of Sigfox.

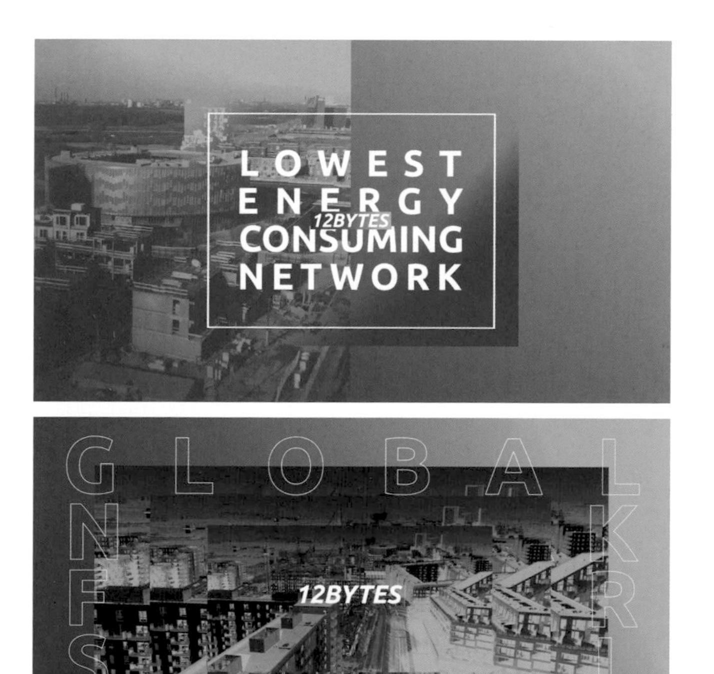

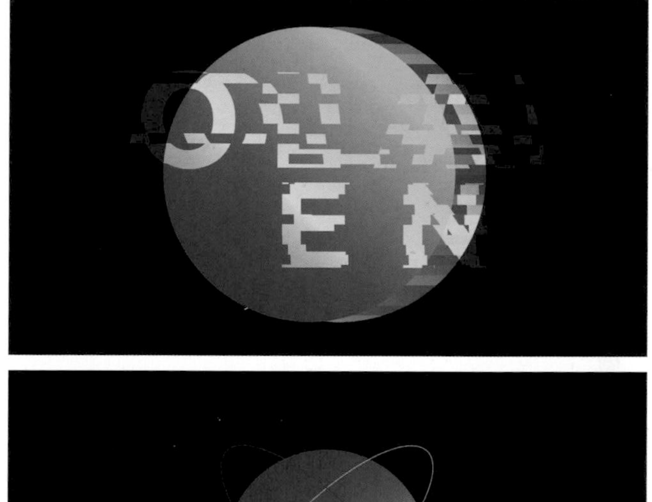

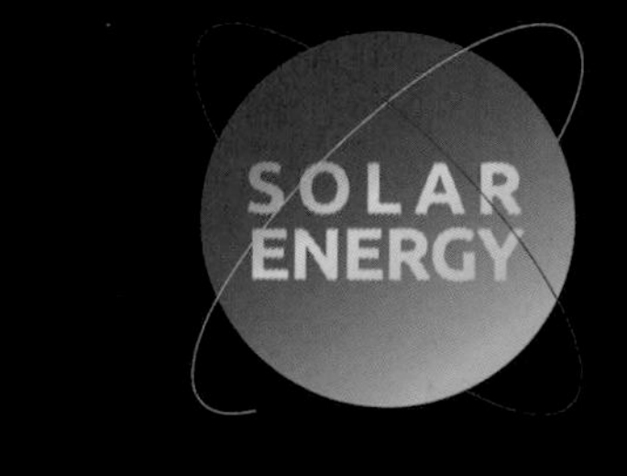

AGENCY **Boogie** / PRODUCTION **Fadereight Films** / CLIENT **Sigfox**

Vimeo

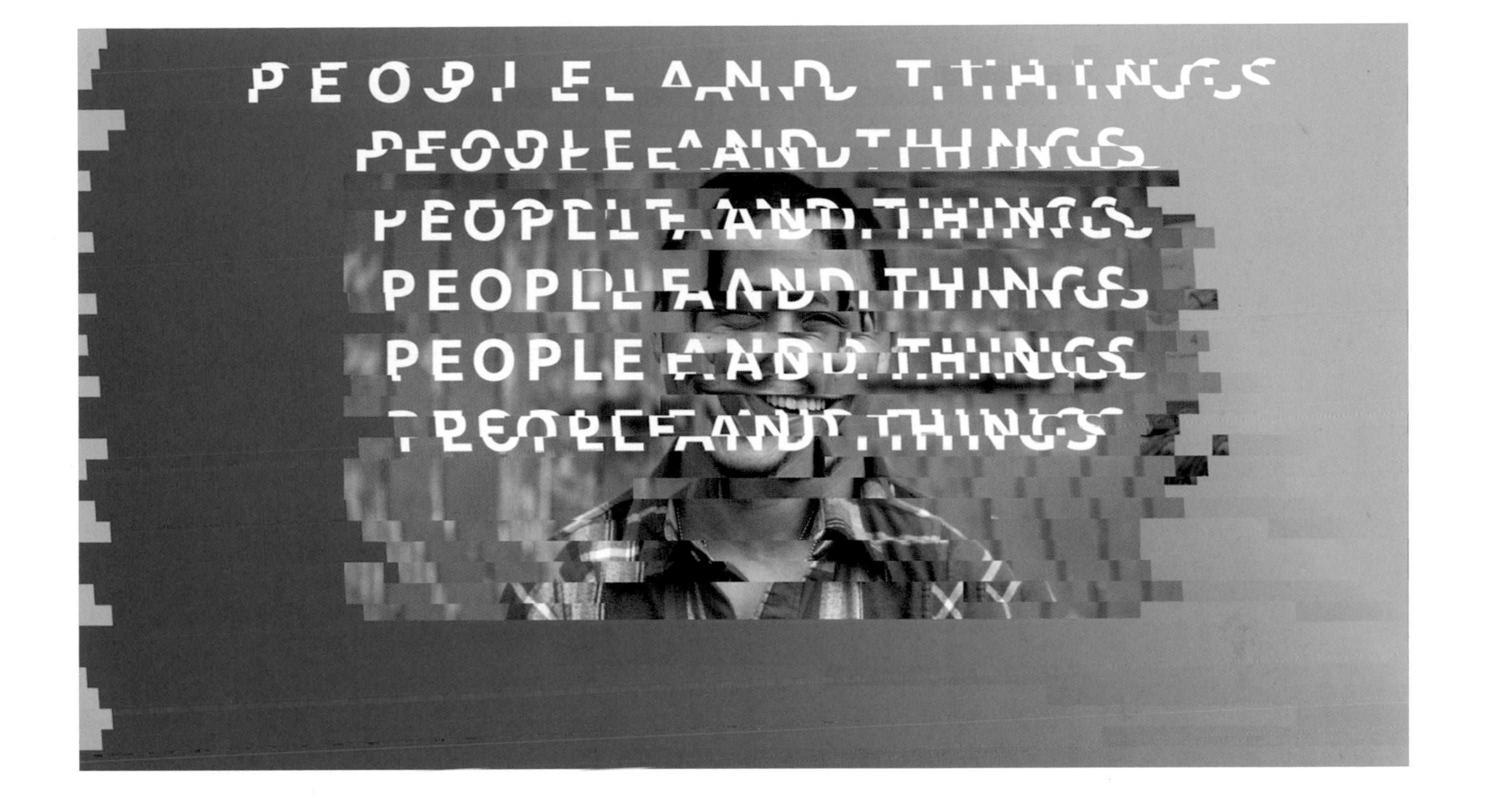
PEOPLE AND THINGS

TERT

MTV EMA 2018 Teaser

BY INLANDSTUDIO

MTV asked INLANDSTUDIO to design the teaser for MTV EMA 2018. Combining with variety of techniques, such as cel animation and 2D animation, INLANDSTUDIO created this dynamite video.

CREATIVE DIRECTION **Gonzalo Nogues** / DESIGN DIRECTION **Javier Bernales**
ANIMATION **Gonzalo Nogues, Javier Bernales, Lucas Gugliara, Julian Nuñez** / CLIENT **MTV**

Official

IBM Think 2019

BY FIELD

Through innovative technology and industry expertise, IBM is changing how the world works by building smarter businesses. Think is IBM's flagship event, inviting business and thought leaders to discuss new and emerging technologies which will fuel the world's next industrial transformation. FIELD designed seven visual systems, each with unique environments and individual expressions of the Think's typography—from highly satisfying motion sequences and smart surfaces to amorphous shapes, chaotic grid systems, and playful characters.

CREATIVE DIRECTION **Marcus Wendt** / STRATEGIC CONSULTANCY **Vera-Maria Glahn**
EXECUTIVE PRODUCTION **Xavier Boivin, Norra Abdul Rahim** / PRODUCTION **Alice Shaughnessy, Joe Smith** / CLIENT **IBM**

Behance
Official

think
think
think

IT & I—An Algorithmic Poem

By OddOne

IT & I is a poem about Peter van de Riet's creative process with software and hardware. To Peter, Ted Hughs' *Crow's Fall* is a very important poem for him and one of the pieces of literature he fed the algorithm. After rearranging the relevant textual content, the final visualization reflects Peter's intuitive process and ambition.

DESIGN **Peter van de Riet** / SOUND **Ejji**

Behance
Vimeo

PSEUDO
MATTER

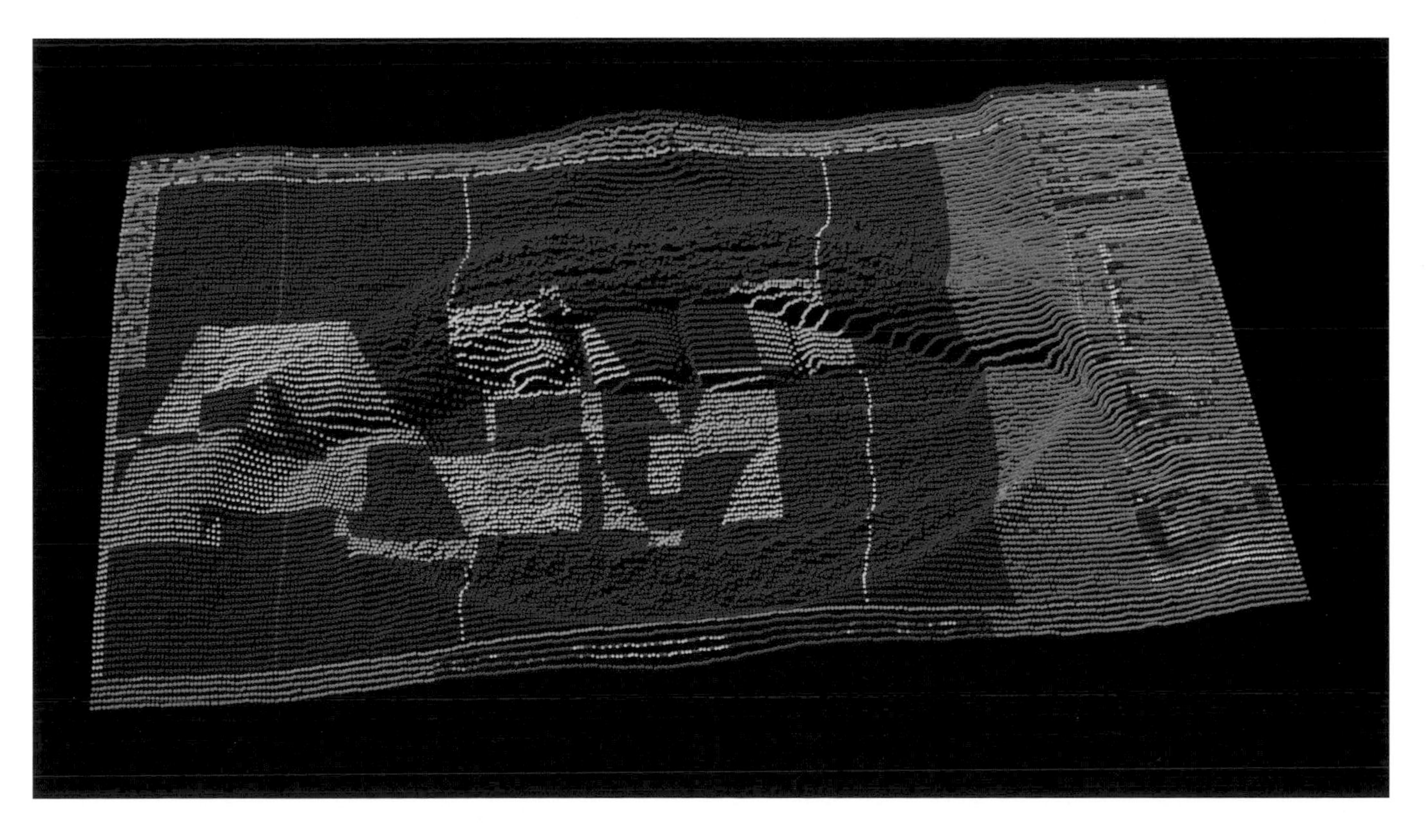

A&E Re-Imagine

BY Manuel Martin

A&E decided to shift to contents with meaning, social impact, grit, and truth. Manuel Martin decided to take some of the themes and iconography from the TV shows and literally transformed them in the motion. Using the chromatic vibration between blue and red, the duotone palette conveys the industrial effect of the A&E's shift to meaning contents.

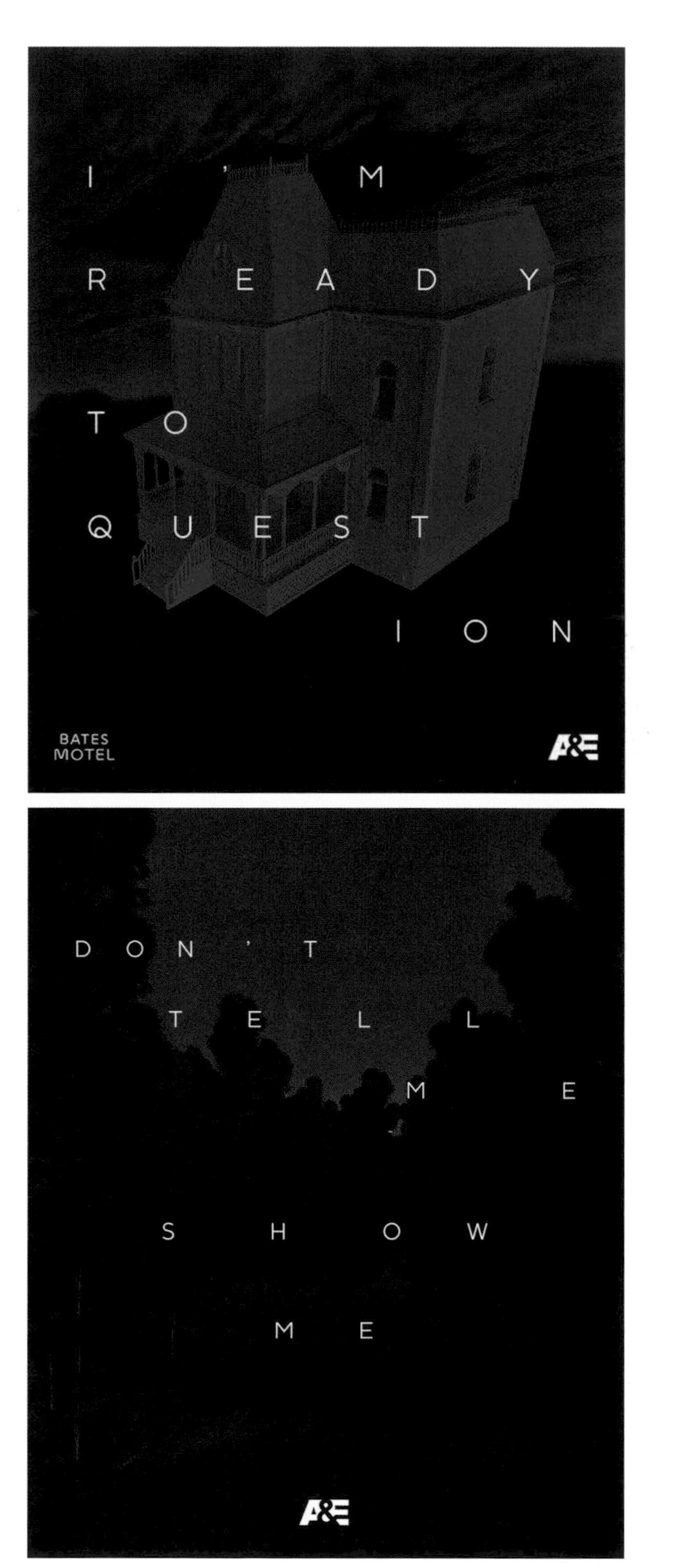

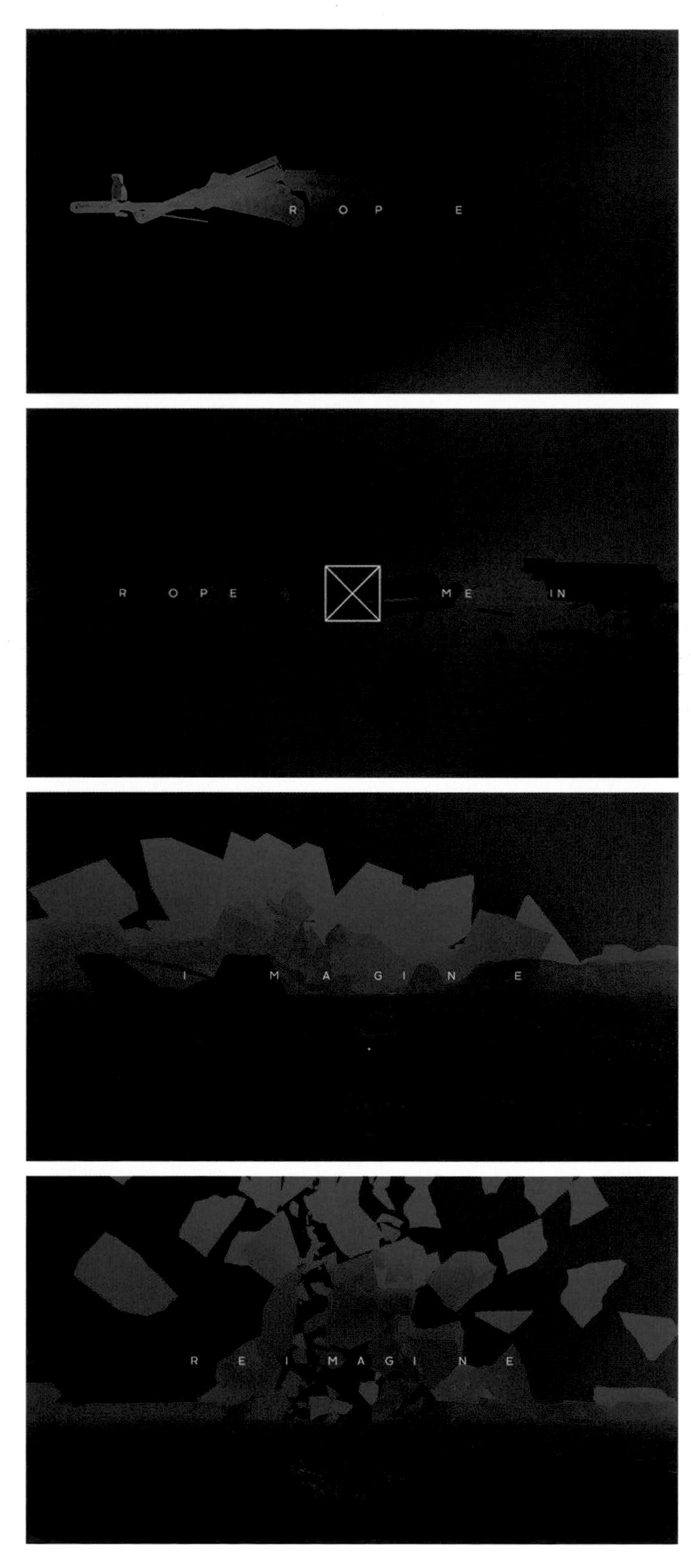

AGENCY **Viewpoint Creative** / CLIENT **A&E**

Ergonofis— Built to Move

BY Sebastien Camden

This promotional video was created to greet the visitors of Ergonofis' new pop-up stores. The designers wanted to create a fast-paced video that would focus on different words expressing what people's body and mind are built to do in the daily work life. A wide variety of techniques such as 2D and 3D animations and stop-motions were used to enhance the concept and make the video as spectacular as possible.

Vimeo

ART DIRECTION **Zacharie Lavertu, Thomas J Salaun** / CREATIVE DIRECTION **Zacharie Lavertu, Sebastien Camden**
PRODUCTION **Evren Boisjoli** / ANIMATION **Sebastien Camden, Aaron Kaufman**

Carmageddon

BY Ranger & Fox

This project is a homage to Los Angeles—the city Ranger & Fox loves. Despite its congested shortcomings, Los Angeles is still a very cool place to live. Ranger & Fox wanted this project to reflect those characteristics—the grinding traffic, the helplessness of perpetual stop-and-go. Meanwhile, those characteristics were transformed into something visually unique highlighting the beauty among the chaos.

SOUND **Echolab**

Official

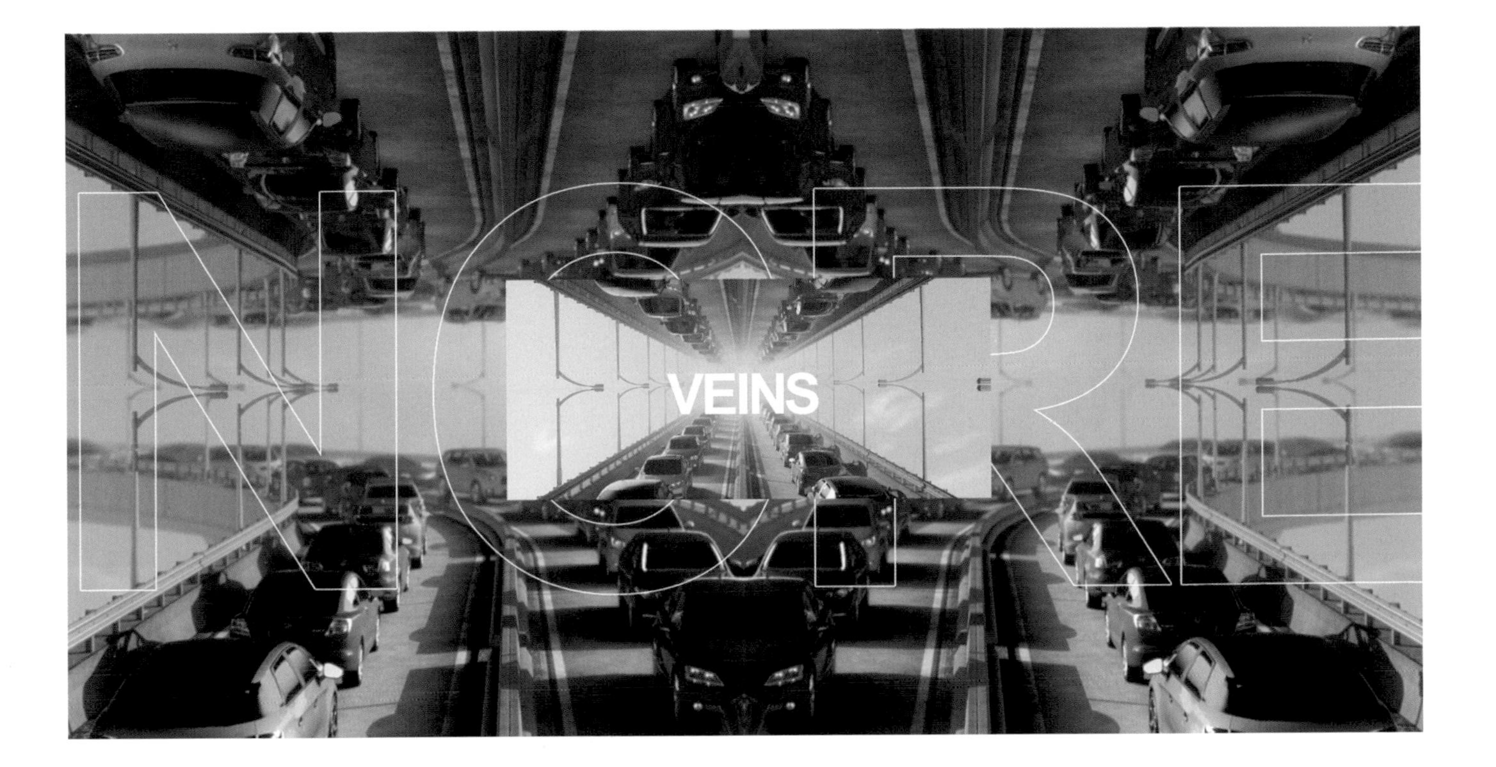
VEINS

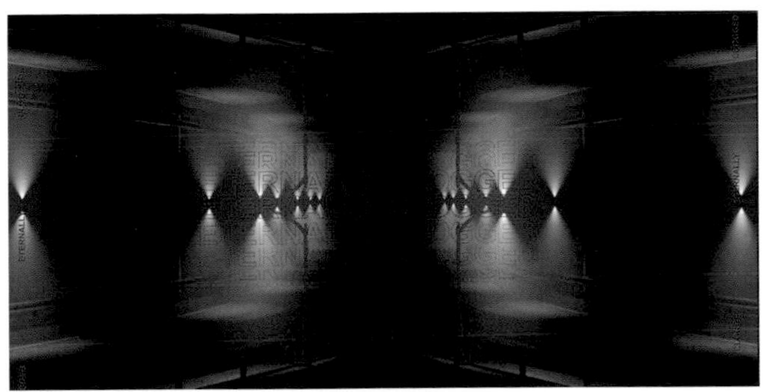

NON NON
O ELOVEL
O LOST L
NO
LOVE
LOST

130
HOURS
PER DRIVER
PER YEAR

Bold Typo Opener

BY Wladyslaw Lutz

This motion graphic project for the modern branding identity was created by eye-catching typographic animations, monochrome palette, and distortion video effects. In Wladyslaw Lutz's opinion, a combination of bold font and minimalistic style can feature a distinct branding concept in a short form.

Vimeo

TYPOGRAPHY

#like
#follow #share
#share
#share
#follow
#follow #like #follow
#like #share
#follow #follow
you

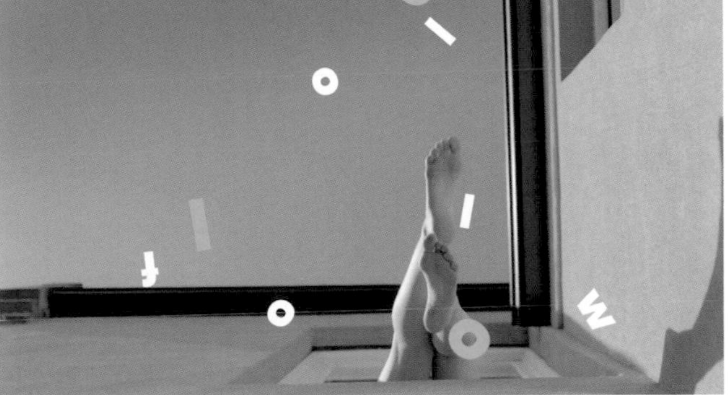

Made in the Middle

BY Ben Radatz

Made in the Middle is a design conference in Kansas City, Missouri, which spotlights artists, makers, and designers from around the Midwest. Ben Radatz collaborated with the event's organisers to adapt their print brandings to motion by developing a process through which clean type can be input and will inherit analogue properties like photocopy stretching, soft edges, and grain.

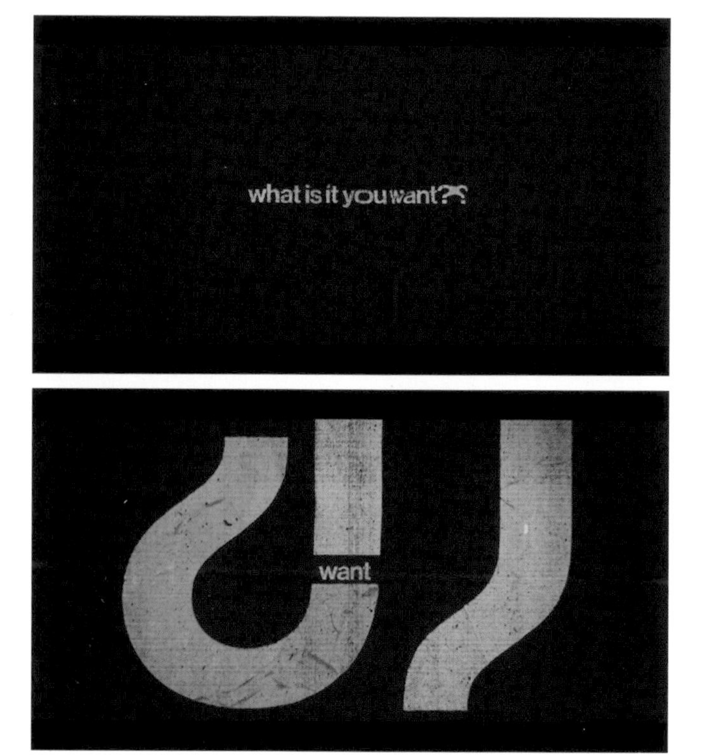

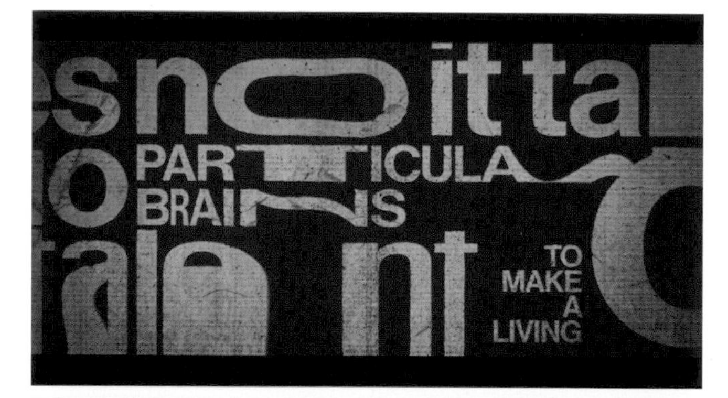

MUSIC **NZCA Lines**

Vimeo

COUNTRY
ANDFRIENDS
FAMILY
LOVEAND
BLEMS
THE
FAMILY
ANDCHILDRENAND
NARROWMINDED
THECOUNTRYPROBL
WORRIES

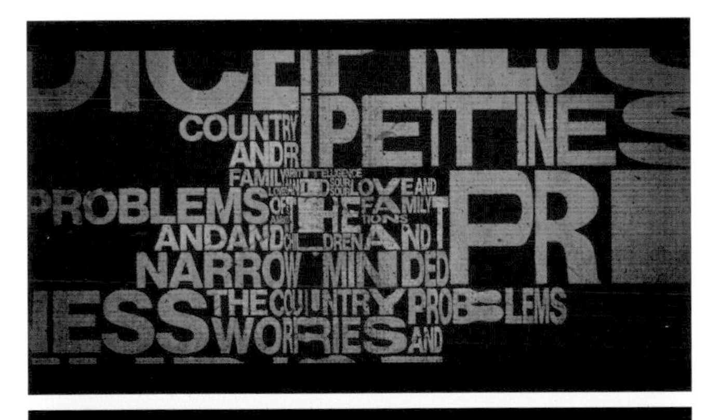
COUNTRY
PETTINES
PROBLEMS
NARROW MINDED
THECOUNTRYPROBLEMS
WORRIES

MADE
IN THE
MIDDLE
USA

OF
ANXIETY

Rodeo Reel

BY Sebastien Camden

This is a manifesto video created for Rodeo—a photo & video production agency. The goal was to create a rhythmic manifesto focusing on typography that would also showcase the work of the studio's roster while delivering the type of energy that the company is all about.

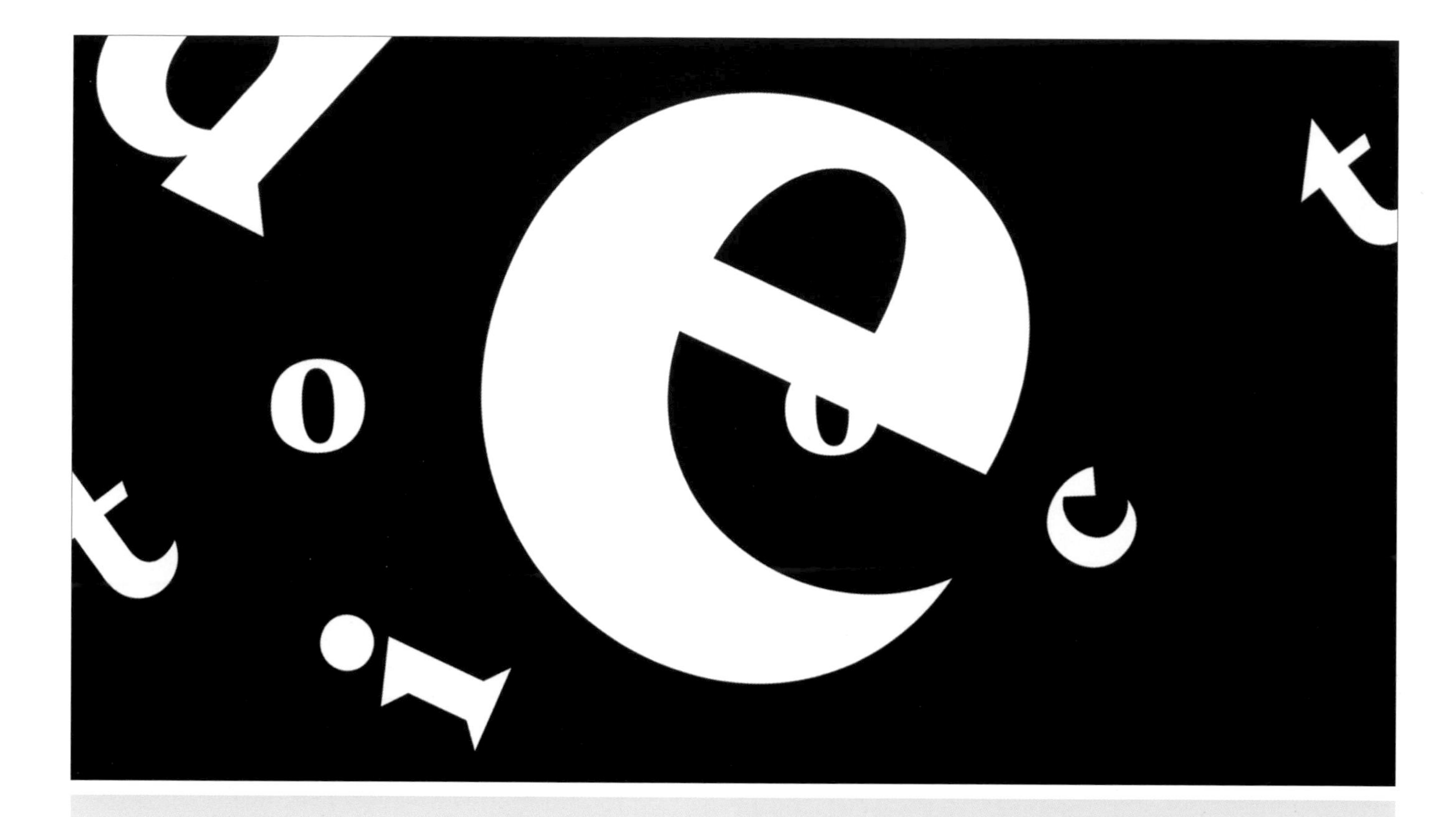

bruno florin
bruno florin
bruno florin
bruno florin
bruno florin
bruno florin
bruno florin

ART DIRECTION **Cossette** / PRODUCTION **Rodeo, Frédérique Quintal** / MOTION DESIGN **Sebastien Camden** / CLIENT **Rodeo**

gabrielle
sykes

DANIEL EHRENWORTH

RAINA &
WILSON

raphaël
ouellet

Festival L*Abore 2019

BY Megan Palero

Festival L*Abore is an annual music festival held in the Campground in Heinsdorfergrund, Germany. During the animation's development, the designers applied a consistent use of typeface that could emphasis the bands and musicians taking part in the festival with bold weights and uppercases. As the logo was designed in advance, the designers integrated its colours and shapes as the main visual to all the sequences. The 2D type does not only create aesthetics but achieves eye-catching effects responsive to different screens and resolutions.

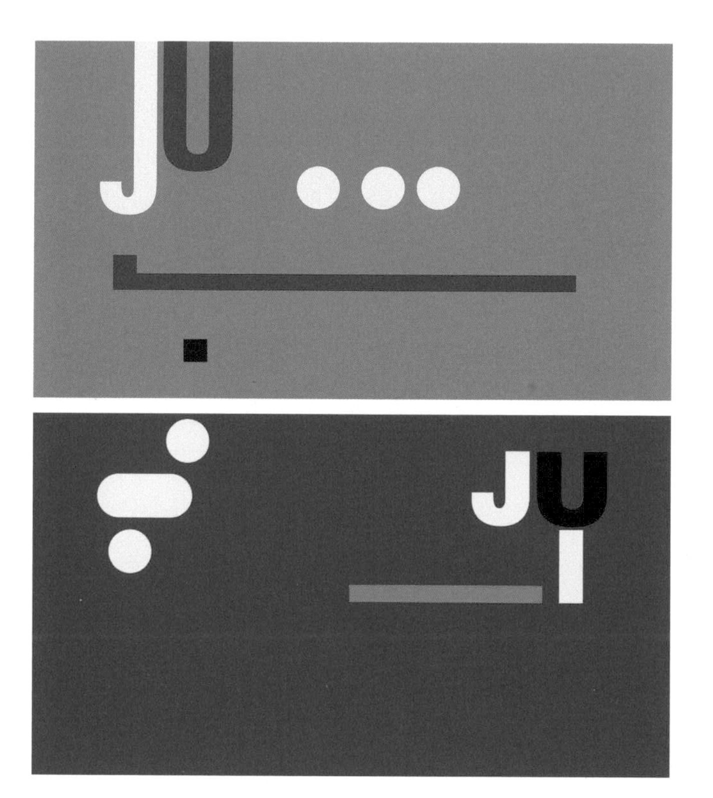

ANIMATION **Megan Palero** / MUSIC **LeRoy** / CLIENT **Festival L*Abore**

Vimeo

L'ABORE
L'ABORE
L'ABORE

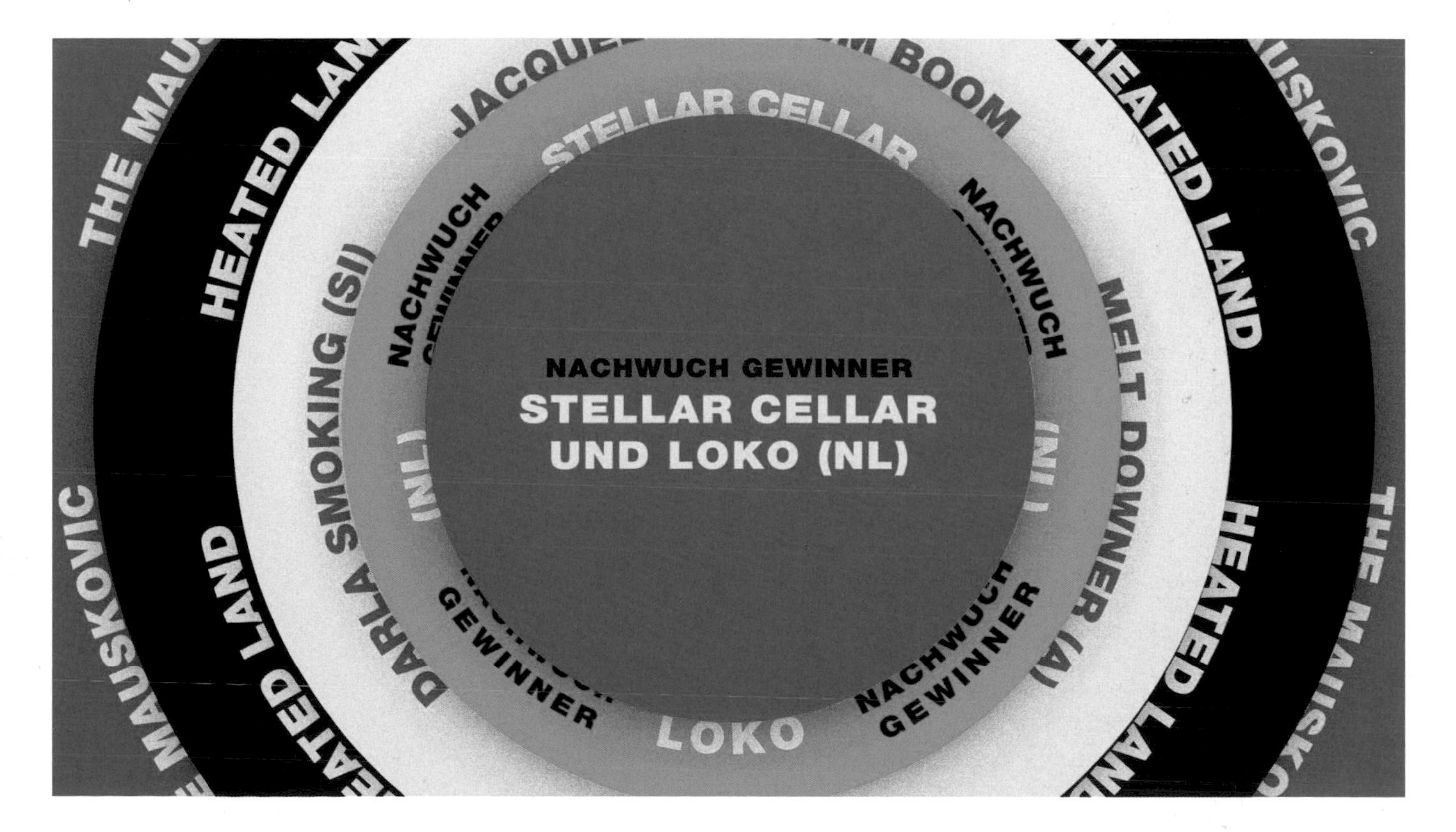
THE MAUSKOVIC
HEATED LAND
BOOM
STELLAR CELLAR
NACHWUCH
DARLA SMOKING (SI)
MELT DOWNER (A)
NACHWUCH GEWINNER
STELLAR CELLAR
UND LOKO (NL)
(NL)
NACHWUCH GEWINNER
GEWINNER
LOKO

It's Always Sunny in Philadelphia S13 Identity for FXX

BY BLOCK & TACKLE

The episodic package for season 13 captures the offbeat Sunny humour with a punk rock edge. To complement this aesthetic, the type needed to be loud and unwieldy. BLOCK & TACKLE created a custom hand-drawn font and added dripping paint animations that embody the wild energy of the show's characters. Iconic Philadelphia imagery, oddball references, and grungy textures round out the design, thrashing together to make people feel like they were just spat out of the centre of a hardcore mosh pit... and cannot wait to dive back in for more.

AGENCY **BLOCK & TACKLE** / CLIENT **FX NETWORKS**

Official

WEDNESDAY
SEPT 5TH

WEDNESDAY
SEASON PREMIERE
TONIGHT

SEASON PREMIERE
SEASON PREMIERE
TONIGHT

La vita è più meglio

BY Bojan Milinkovic

"La vita è più meglio" means "Life is better" in English. This animation is a collaboration between Bojan Milinkovic and his friend Tommaso Lavagnoli. Being inspired by the La Effe's rebranding by Nerdo Studio, Bojan and Tommaso aimed to create a hyperactive animation without pauses. Combing Tommaso's editorial and typographic skills and Bojan's tricks of motion graphics, the animation exhibits flash and dynamics with fast movement and explosive release of energy.

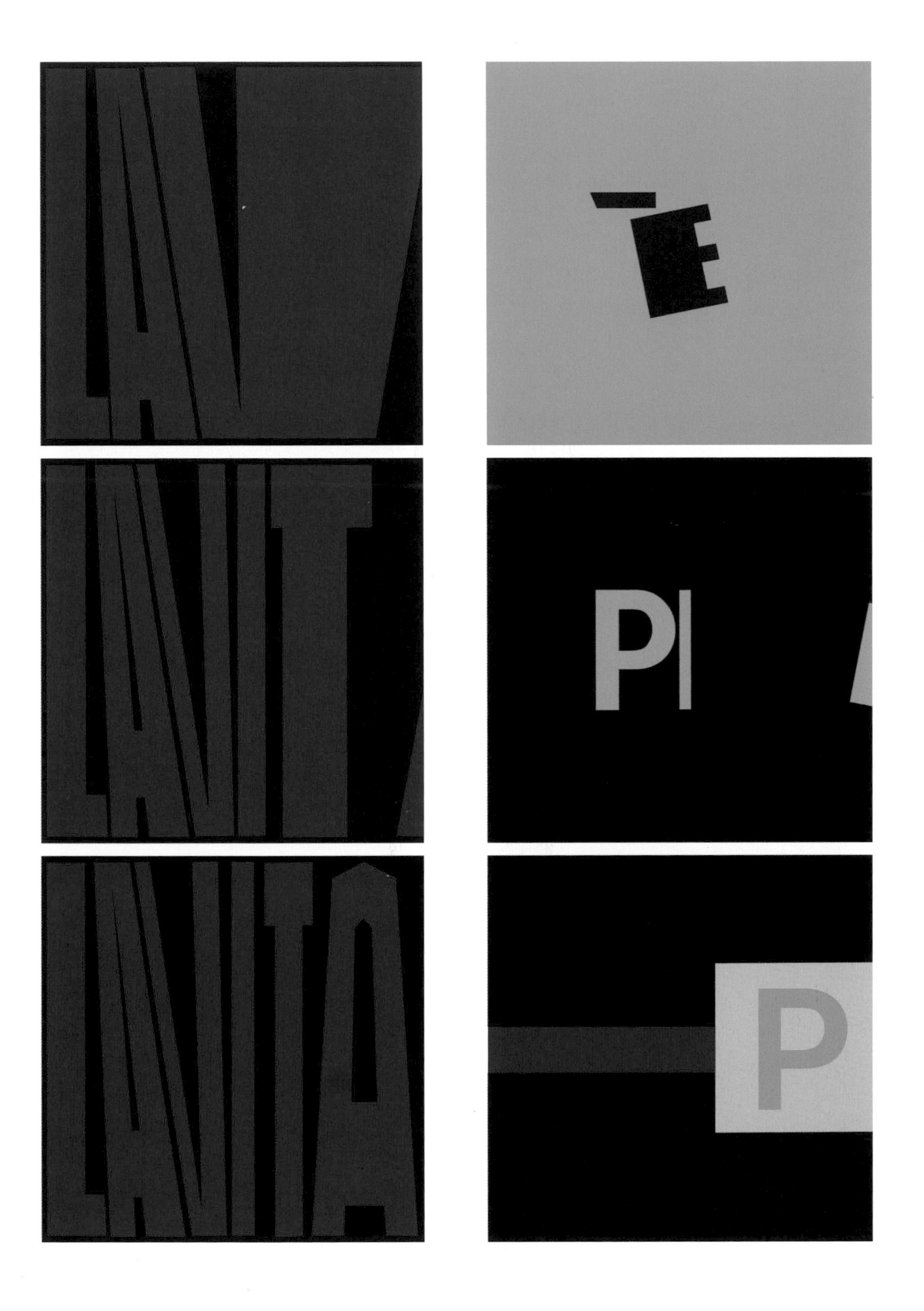

COOPERATION **Tommaso Lavagnoli**

Dribbble

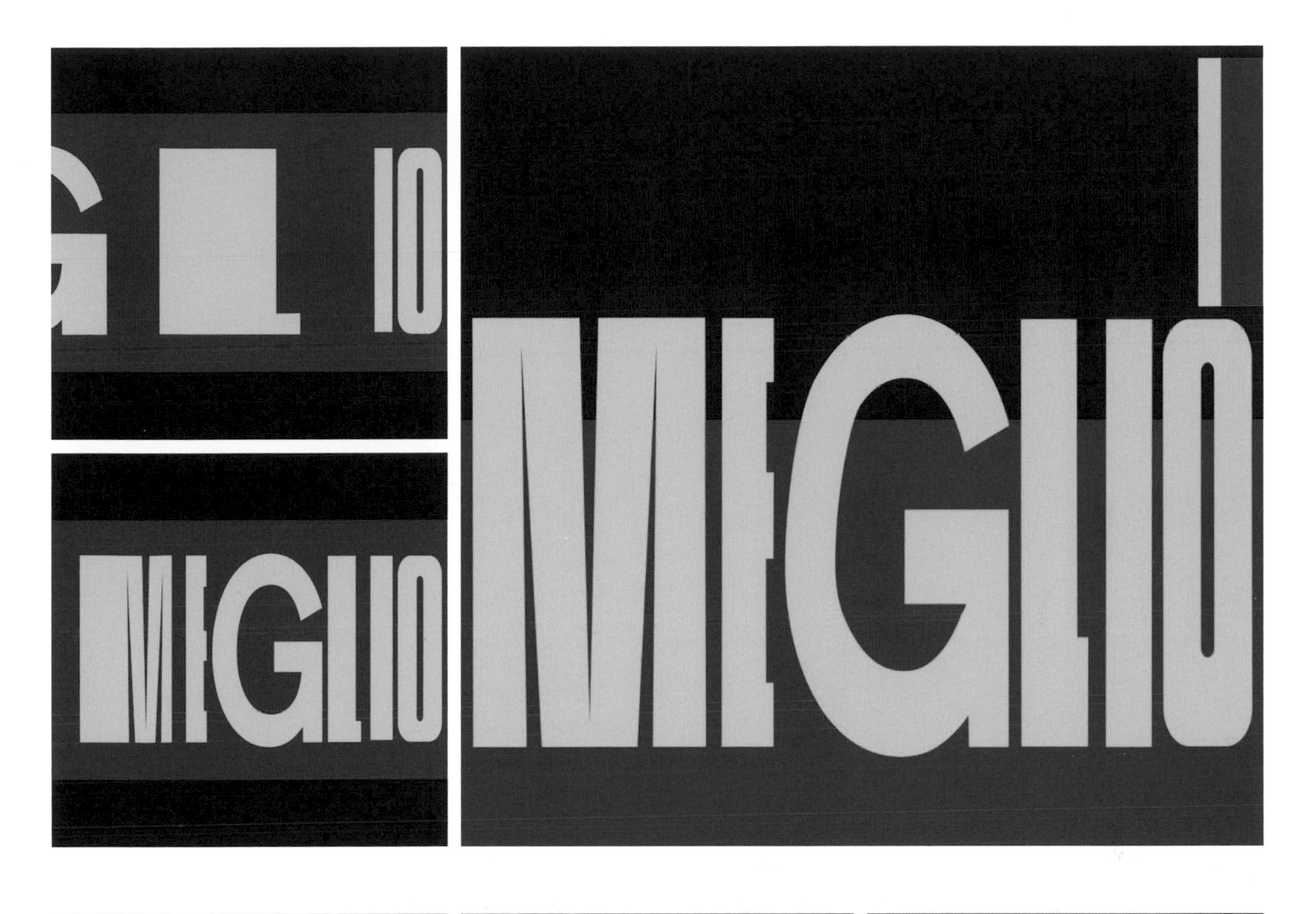

RAD Launch Video

BY Sebastien Camden

This launch video for the journalistic laboratory of CBC—Canada's national television revolves almost entirely around the use of a bold colour palette and a manifesto told by striking words and sentences that aim to intrigue the audience. A well-balanced rhythm is important for the audience to understand what RAD is about while also be thrilling.

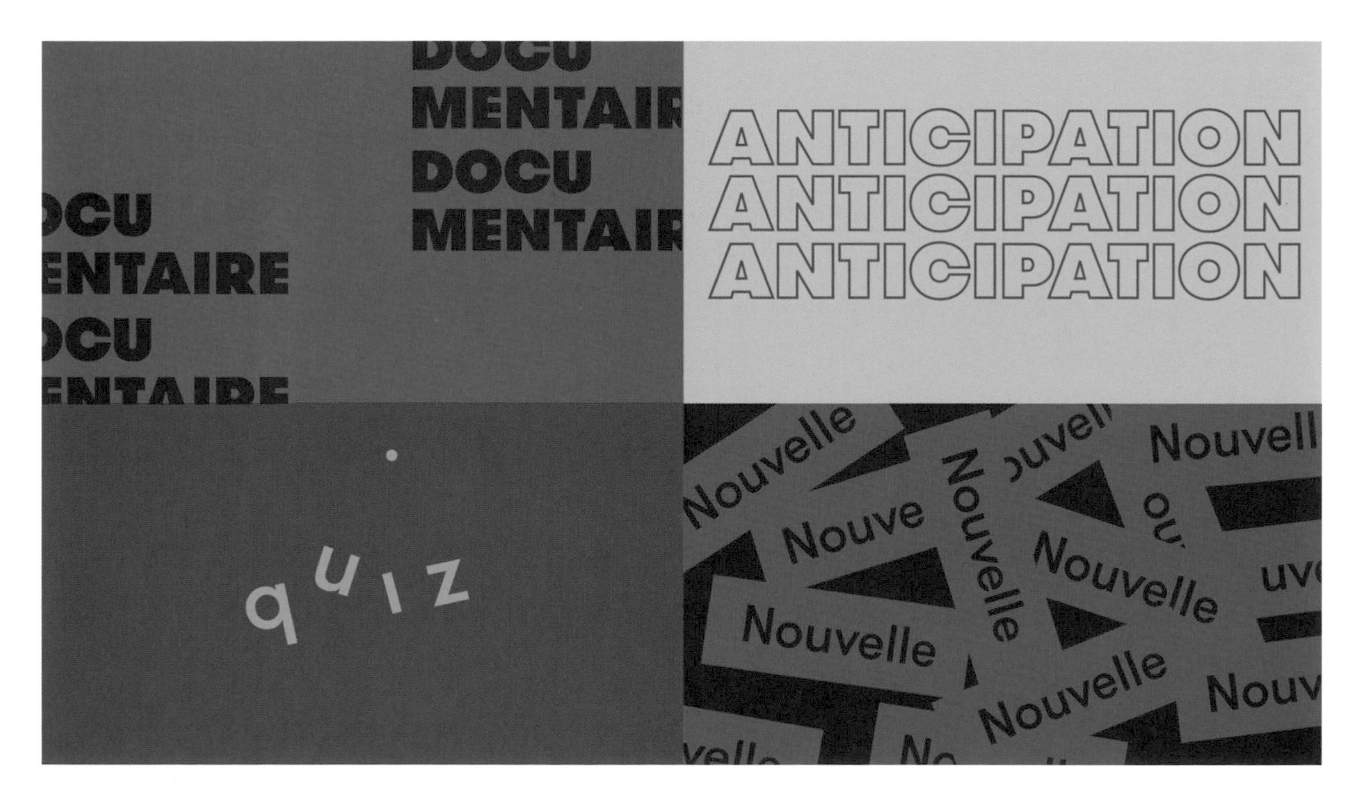

ART DIRECTION **RAD, Charles Fortier** / MOTION DESIGN **Sebastien Camden, Jordan Coehlo (Barth)**
COPYWRITING **RAD, Gigi Huynh** / CLIENT **RAD**

Vimeo

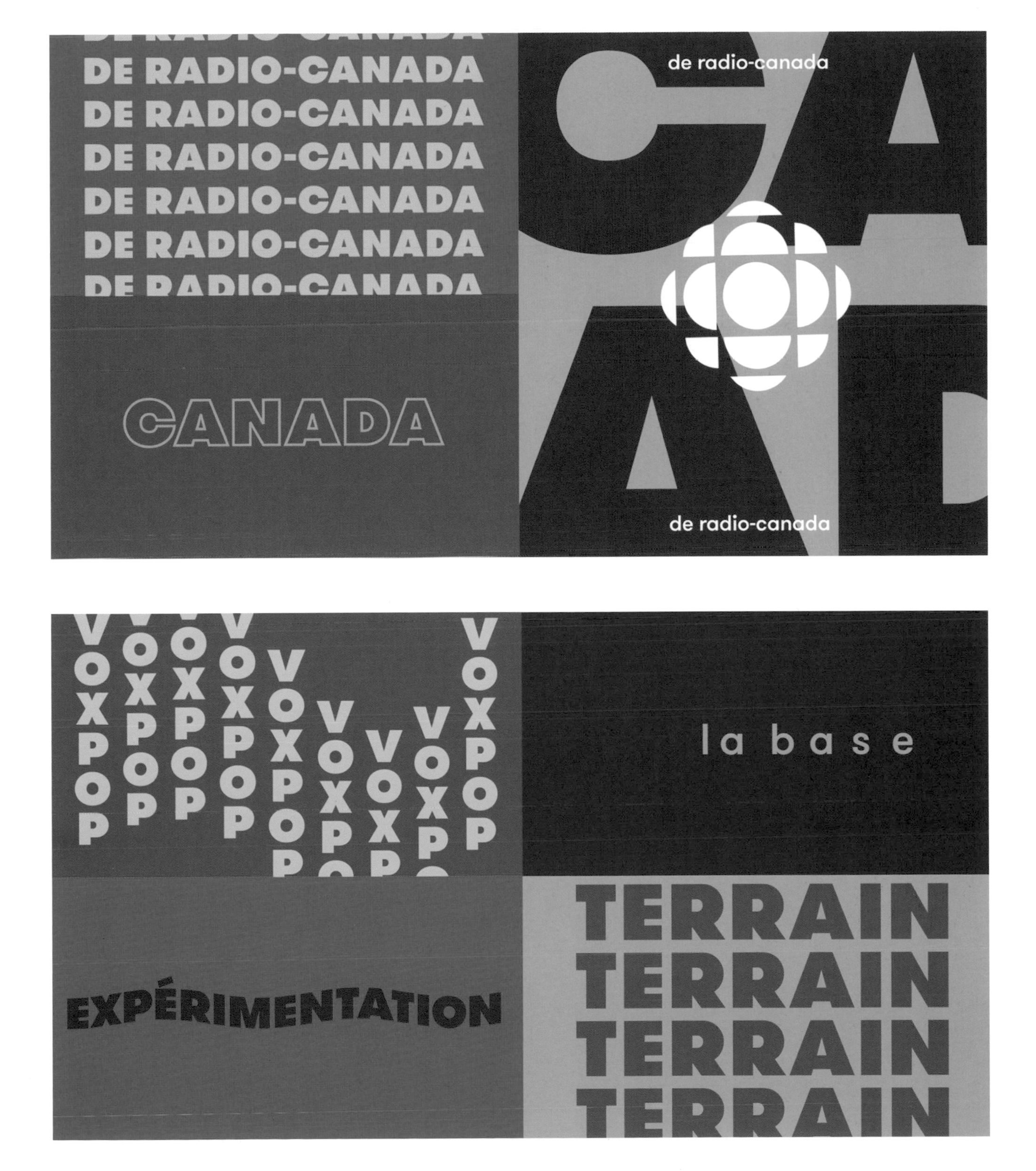
DE RADIO-CANADA
DE RADIO-CANADA
DE RADIO-CANADA
DE RADIO-CANADA
DE RADIO-CANADA
de radio-canada
CANADA
de radio-canada
VOXPOP
la base
EXPÉRIMENTATION
TERRAIN
TERRAIN
TERRAIN
TERRAIN

Buck the Vote

BY Buck

This video was created as a part of the series called #buckthevote. Utilising the voice of Buck, the designers wanted to encourage people to turn out and vote. The dynamic typography and shifting layouts, along with colour changes, allows the viewers to draw attention to various phrases and important information, such as the election's date and #buckthevote hashtag.

Vimeo

CREATIVE DIRECTION **Orion Tait, Daniel Oeffinger, Steve Day** / PRODUCTION **Alexi Yeldezian, Kirsten Collabolletta**
DESIGN **Yker Moreno** / ANIMATION **Joe Brooks**

WEBPAGE

Cuchillo Website

BY Cuchillo

Cuchillo considers that the best way to showcase the website projects is to avoid mock-ups and case studies and get back to basics. Since images and texts are not necessary, Cuchillo purely opens a window for the visitors to explore every project in the way as conceived—a slick design "abusing" of gigantism with a touch of brutalism. In the portfolio website of Cuchillo the animations and transitions in the style of *Pixel Perfect* offer fluidity and dynamism. And the interactive contact form adds a quirky finishing touch.

CREATIVE DIRECTION **Rober Alonso Maide** / INTERACTIVE DIRECTION **Javier Corrales** / TECHNICAL DIRECTION **Alberto Bajo**
DESIGN **Adrian Romero, Juan Saiz, Naiara Borderia** / MANAGEMENT **Maarten Volckaert**

Official

azab info projects zoom news contact
01/02
TELL US SOME??
PROJECTS
WORKS
AZAB
cuchillo
EO!!
Cuéntanos qué
##@%*@ quieres
Estamos en Conde mirasol 11, 48003 Bilbao, puedes pasarte cuando quieras, mandarnos un mail a info@somoscuchillo.com o si eso no te vale te ayudamos para que no tengas que escribir nada. Bueno, tu email sí que lo necesitamos, hasta ahí no llegamos...
¿Quién eres?
¿Qué quieres hacer?
CUCHILLO IS A CREATIVE AND INDEPENDENT GANG
PREV NEXT WORKS ABOUT CONTACT

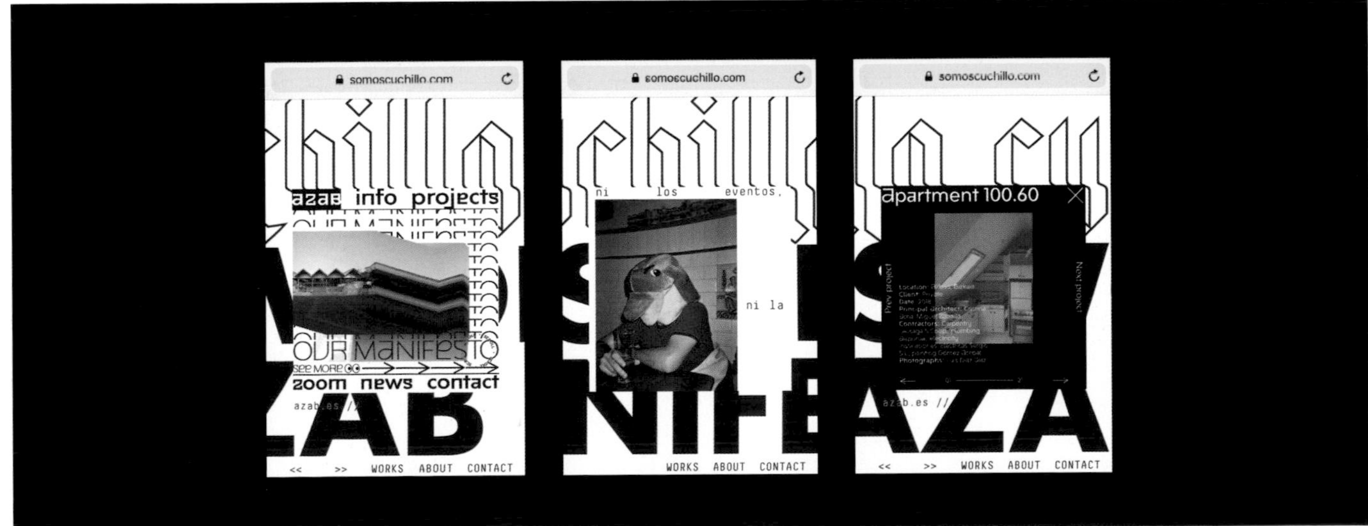
somoscuchillo.com
azab info projects
OUR MANIFESTO
see MORE
zoom news contact
azab.es //
ni los eventos.
ni la
apartment 100.60
Prev project
Next project
<< >> WORKS ABOUT CONTACT

AZAB Website

BY Cuchillo

This project is a part of a complete corporate identity for AZAB, entirely developed by Cuchillo. Cuchillo broke the rules of UX/UI on purpose, designed without margins, repeatedly added overlapping elements, and finished the design all with a touch of gigantism. Meanwhile, Cuchillo attached importance to the website's details, such as the infinite gallery with the glitch effect in the style of Win98 and the wavy paragraph in the info section. Cuchillo considers that one of the biggest challenges for this project was to create brutalist and eye-catching transitions without losing the smoothness and fluidity of all transitions and animations.

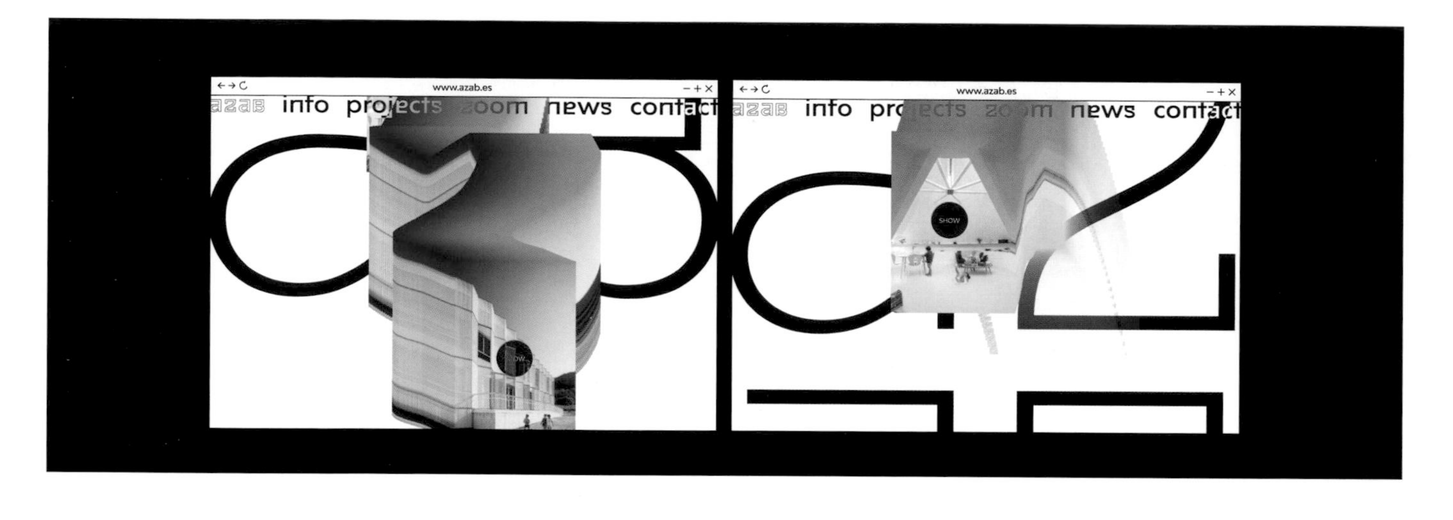

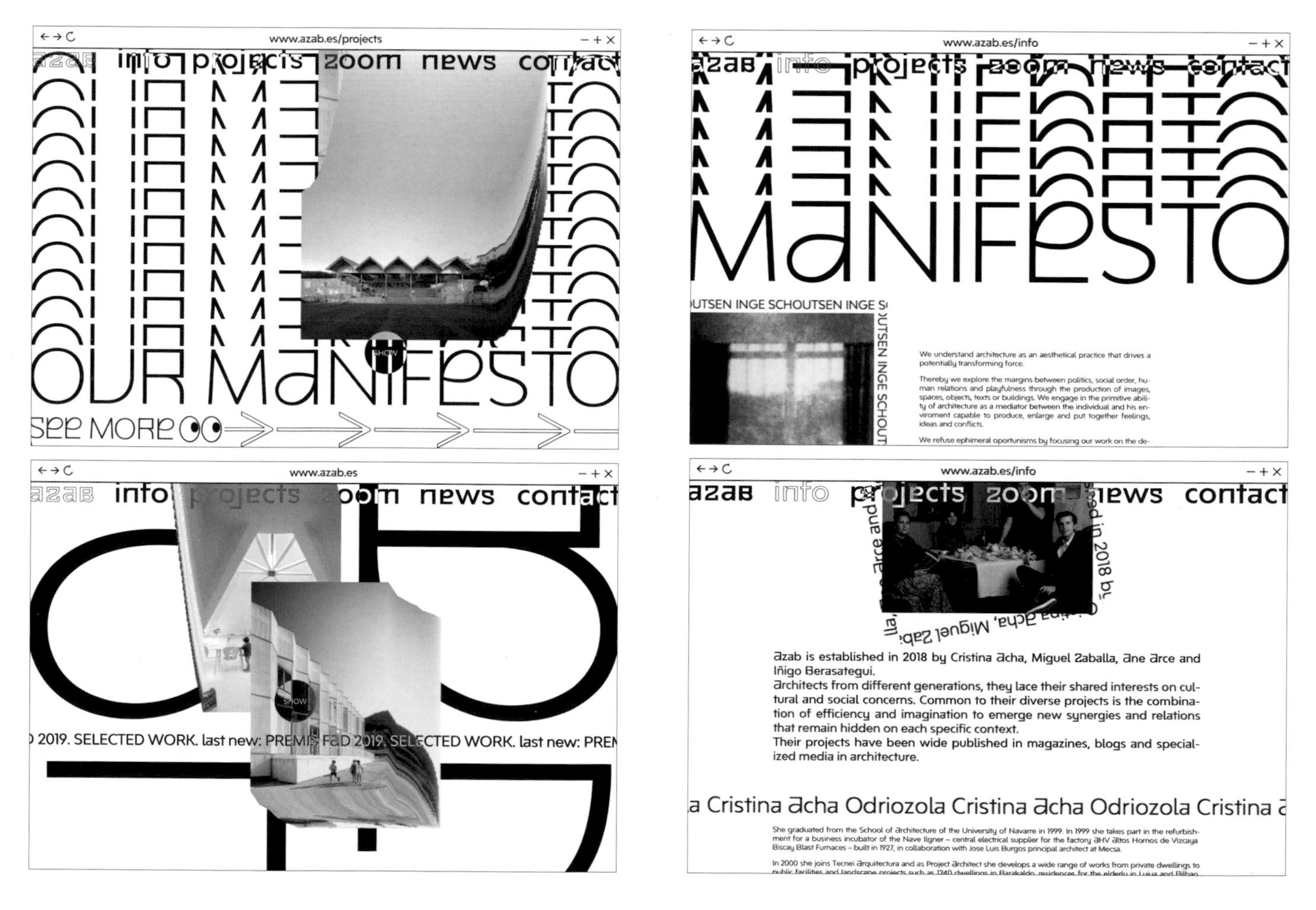

CREATIVE DIRECTION **Rober Alonso Maide** / INTERACTIVE DIRECTION **Javier Corrales** / TECHNICAL DIRECTION **Alberto Bajo**
DESIGN **Adrian Romero, Juan Saiz, Naiara Borderia** / MANAGEMENT **Maarten Volckaert**

Official

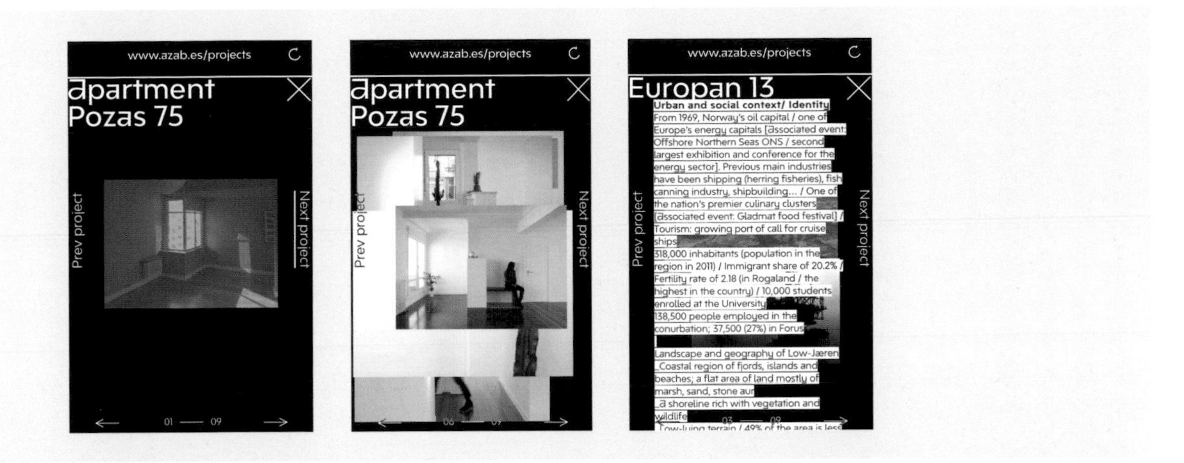
www.azab.es/projects
apartment Pozas 75
Prev project
Next project
01 — 09
Europan 13
Urban and social context/ Identity
From 1969, Norway's oil capital / one of Europe's energy capitals

www.azab.es
azab info projects
WORK. last new:
REMIS FAD 2019. SE
OUR MANIFESTO
SEE MORE
zoom news contact
www.azab.es/projects
PROJECTS
NAME
YEAR
001 Kindergaten extension altzaga 2019
002 Primary School extension Zubiz 2019
003 Herriko Plaza 2019
004 apartment 100.60 2019
005 35 public dwellings "el campillo 2019
006 City seeks architecture 2018
007 aurregoiti Etxea 2018
008 5 public dwellings "Concepción 2018
009 apartment Ribera 2018
www.azab.es/contact
Sotera de la Mier 8
48929 Portugalete
Bizkaia,Spain
+34 944 725 734
info@azab.es
@azab_architects
OUR THOUGHTS
Terms of service
Cookies Policy

URSA MAJOR SUPERCLUSTER

BY Daniel Spatzek

In order to catch the users' attention, Daniel Spatzek decided to create a playful and unique experience for the website entrance without being weird or awkward. Daniel came up with the idea of kinetic diagonal lines that can follow the movements of the cursor. And the navigation with kinetic typography locates on the certain points on the index page.

Behance

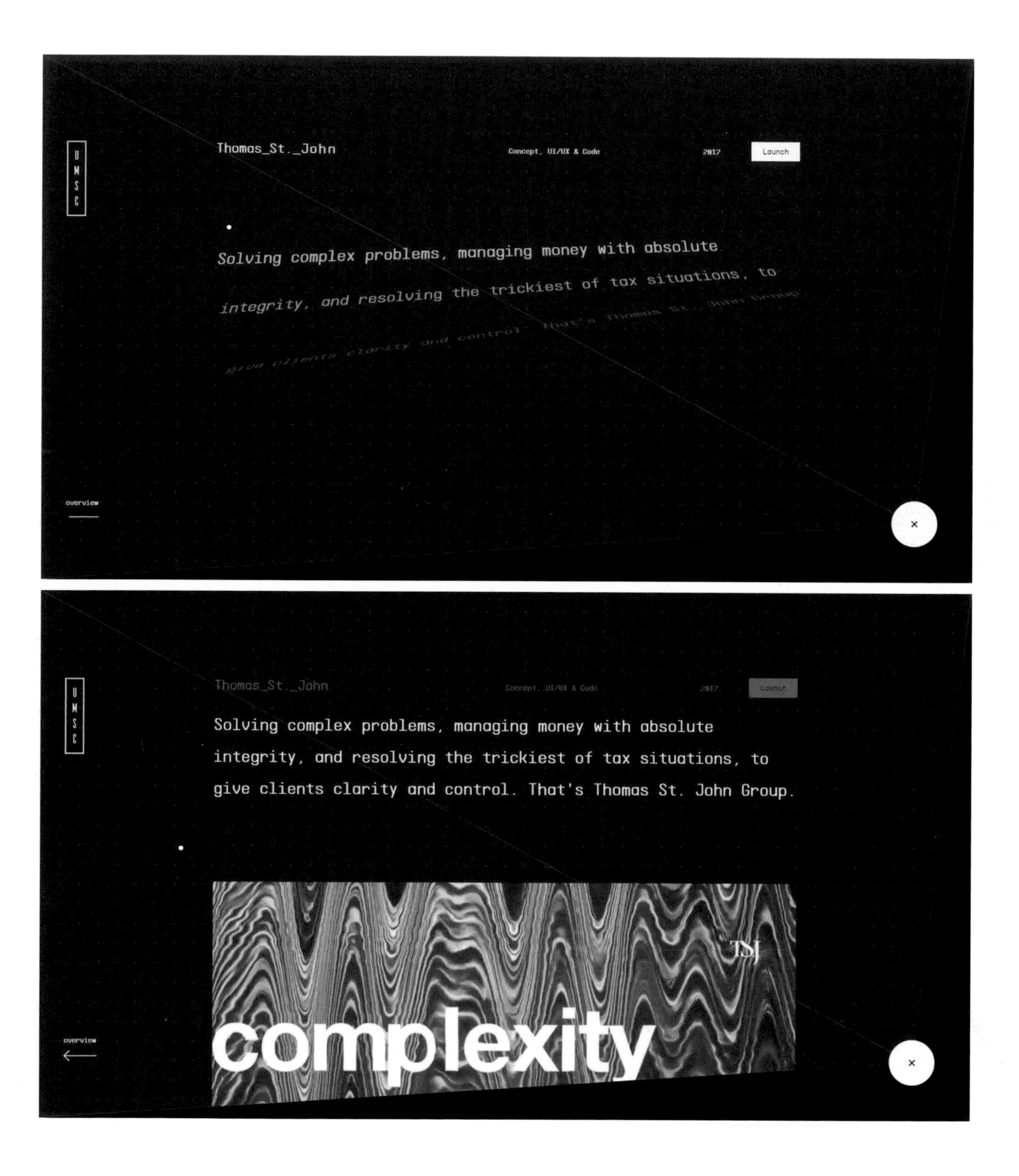

CLIENT **URSA MAJOR SUPERCLUSTER**

WEBPAGE

naked Hub Web Design Concept

BY Bezantee Bao

For the redesign of naked Hub website, Beazntee Bao utilised the generality of circle shape to convey the visual concept of clock and timeline, with which the page transition can be smoothly built. Besides, the visual conflict of serif and sans-serif will help users to distinguish the primary elements and secondary ones hierarchically. Enough breathing space is left on the homepage to give users a better understanding of the detail of the website.

Behance

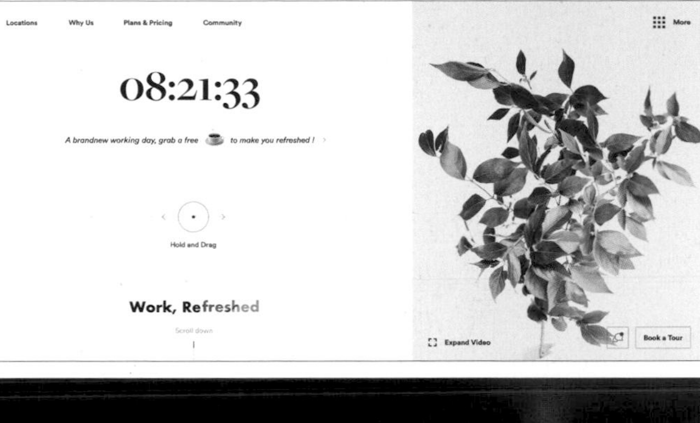

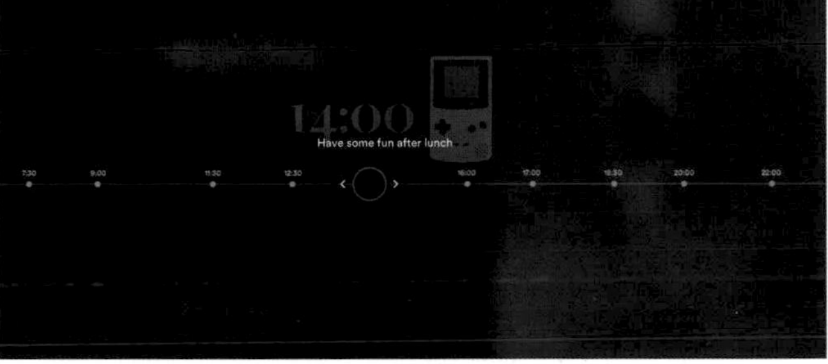

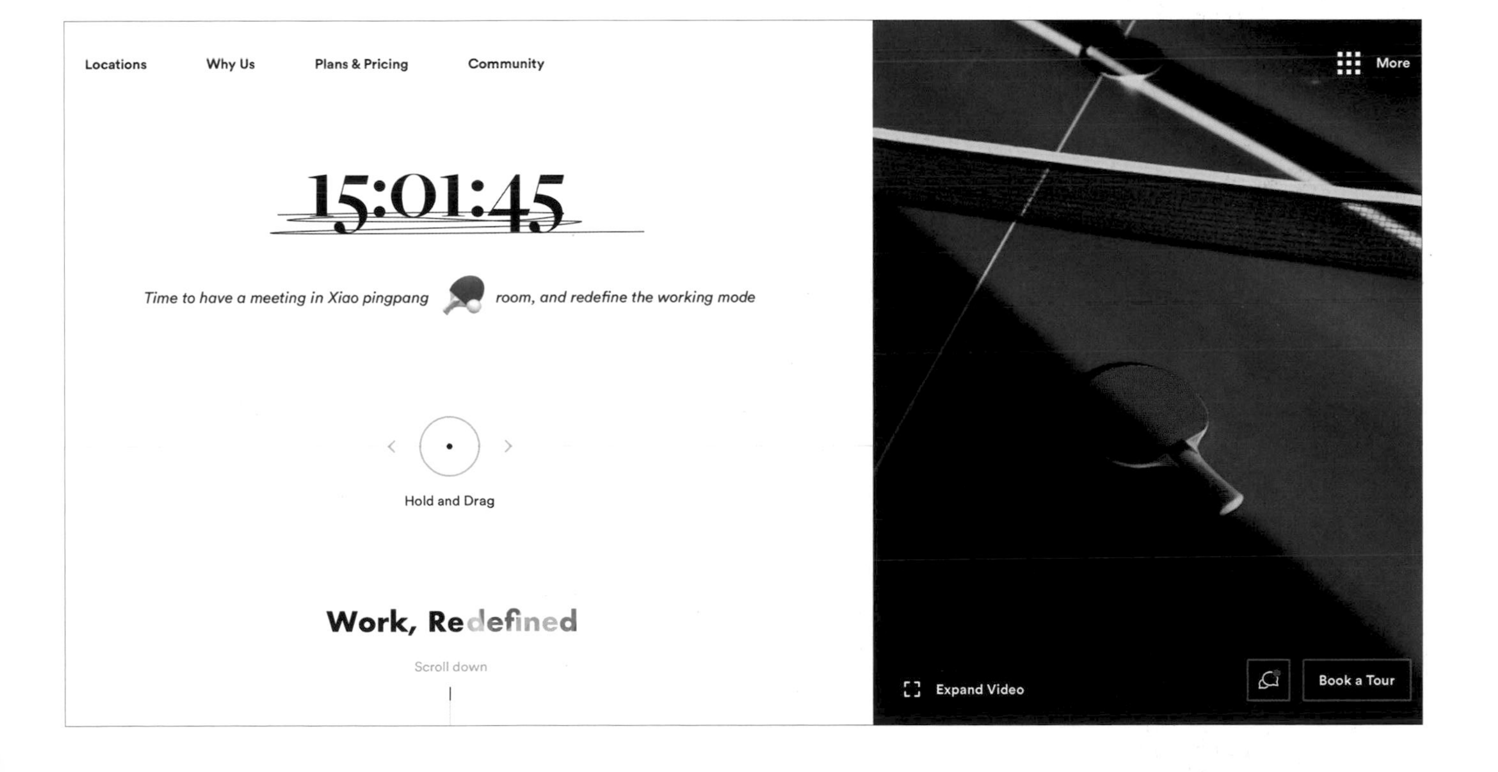

ART DIRECTION **Bezantee Bao** / CLIENT **naked Hub**

Union Merch Website Design

BY Manuel Rovira

In this project, Manuel Rovira tried a more minimalist style with the use of a white interface that naturally cost him lots of time to explore since he needed to change his common style with dark colours. Meanwhile, Manuel was asked to display something in swag style. So Manuel designed various dynamic typographic motions on the interface to fit the minimalist style and come into a feeling of shocking touch.

CLIENT **Union Merch**

Behance

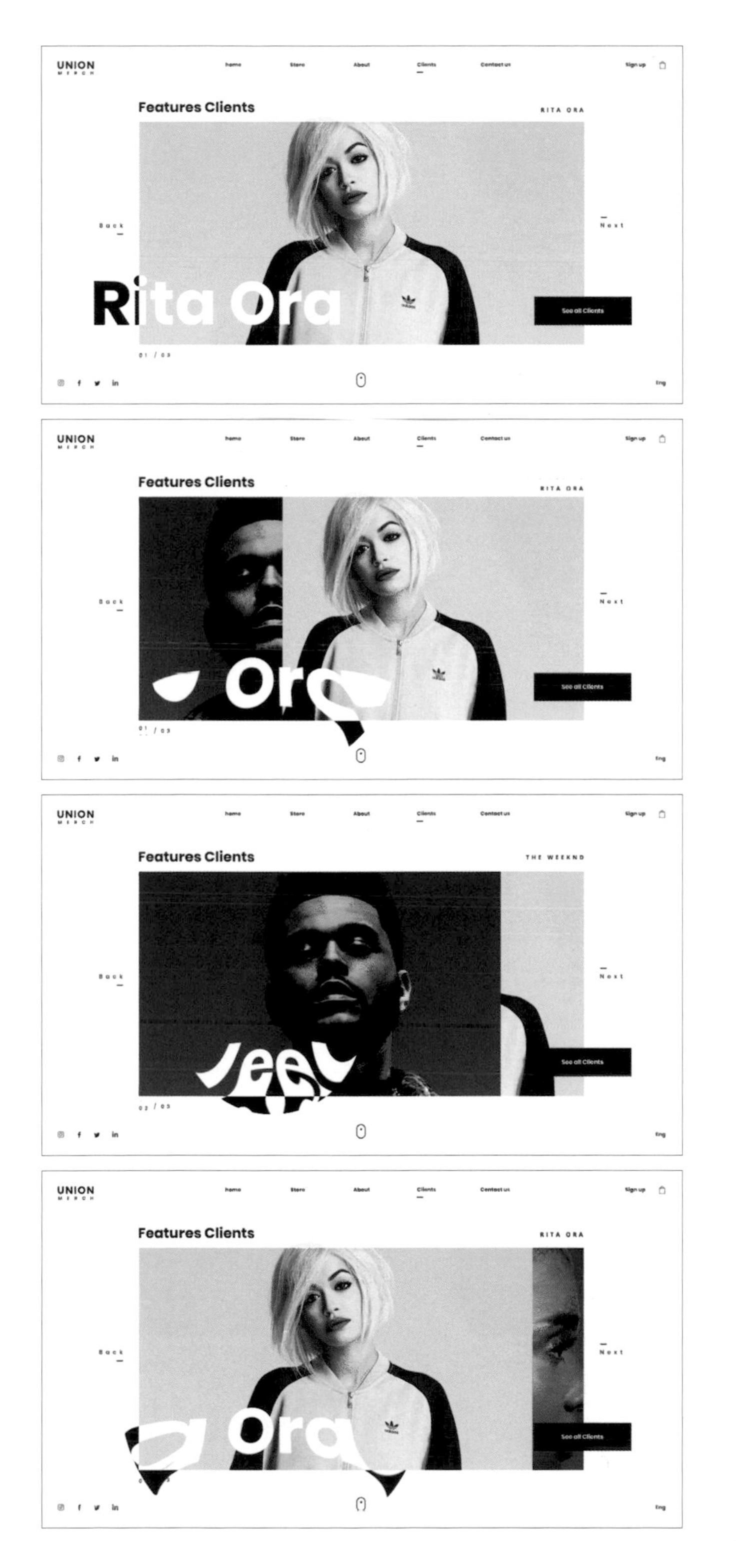
UNION
Features Clients
RITA ORA
Rita Ora
THE WEEKND

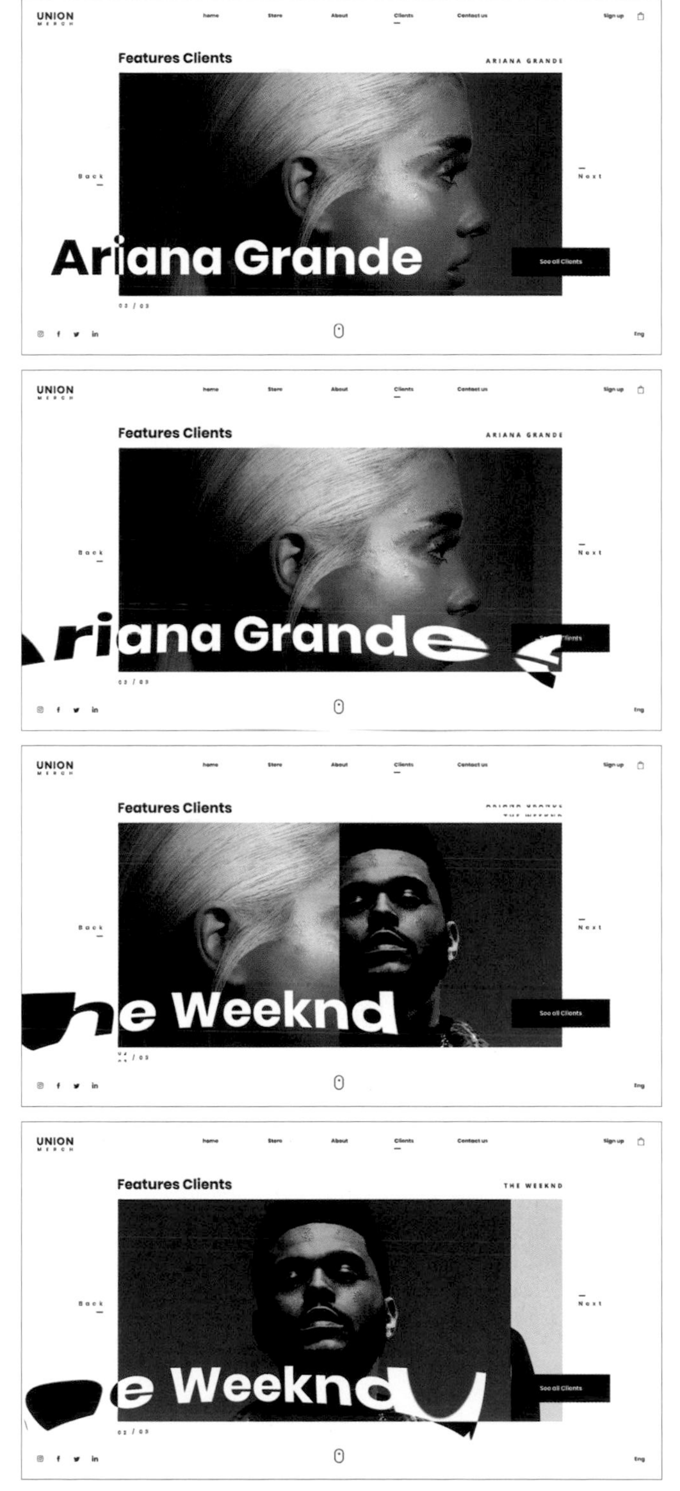
UNION
Features Clients
ARIANA GRANDE
Ariana Grande
e Weeknd
THE WEEKND

Anthony Acosta Portfolio

BY Cody Cano

Anthony Acosta is a prolific and widely celebrated skateboarding/lifestyle photographer based in Los Angeles. Inspired by the urban environments and Los Angeles' unique aesthetics, this website combines fluid animations and a dynamic navigation system with a stark and minimalist design to reflect Anthony's controlled approach to the rawness and irreverence of his subjects.

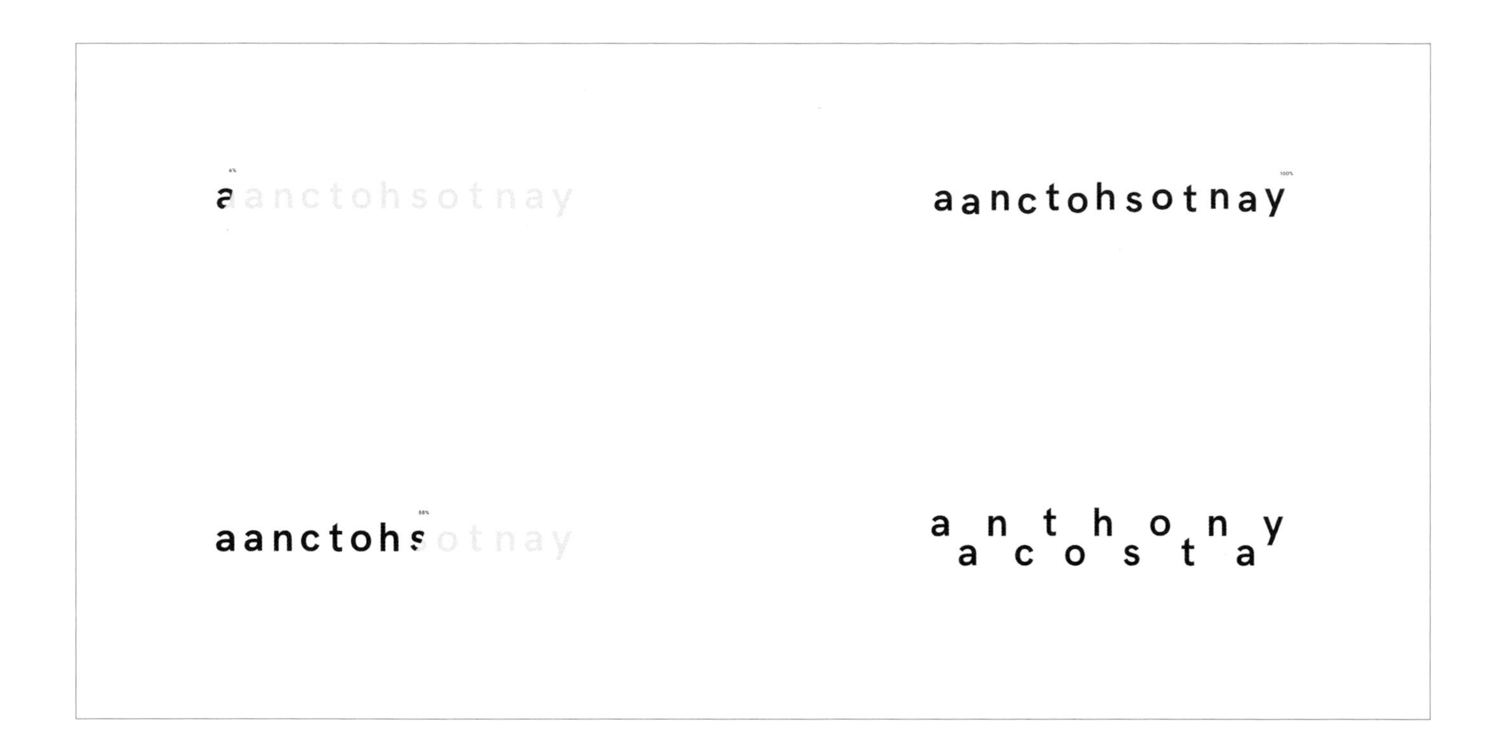

CLIENT **Anthony Acosta**

WEBPAGE

Behance

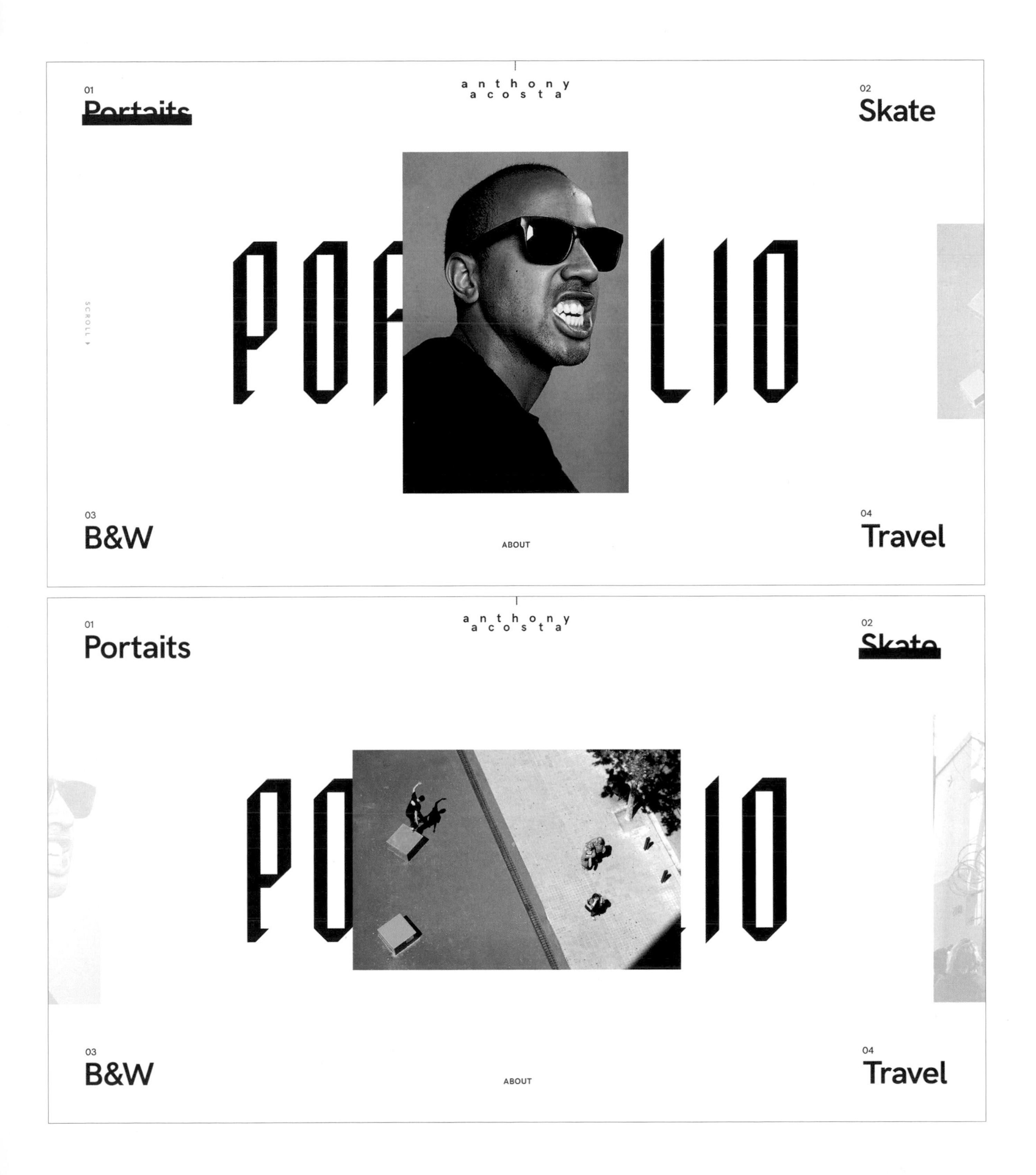

01
Portaits
anthony
acosta
02
Skate
03
B&W
ABOUT
04
Travel
01
Portaits
anthony
acosta
02
Skate
03
B&W
ABOUT
04
Travel

Daniel Spatzek Portfolio 2018

BY Daniel Spatzek

The main idea of Daniel Spatzek was to create a portfolio that focuses on the design content but features the minimalist aesthetic with unique appearance and feeling. To achieve this goal, Daniel chose Suisse Int'l Book and Helvetica Neue LT Pro as the web fonts and employed the 12-column grid layout for the website. The text slice effect has been added to provide the landing page with a more distinctive look and direct the users' eyes to Daniel's own projects.

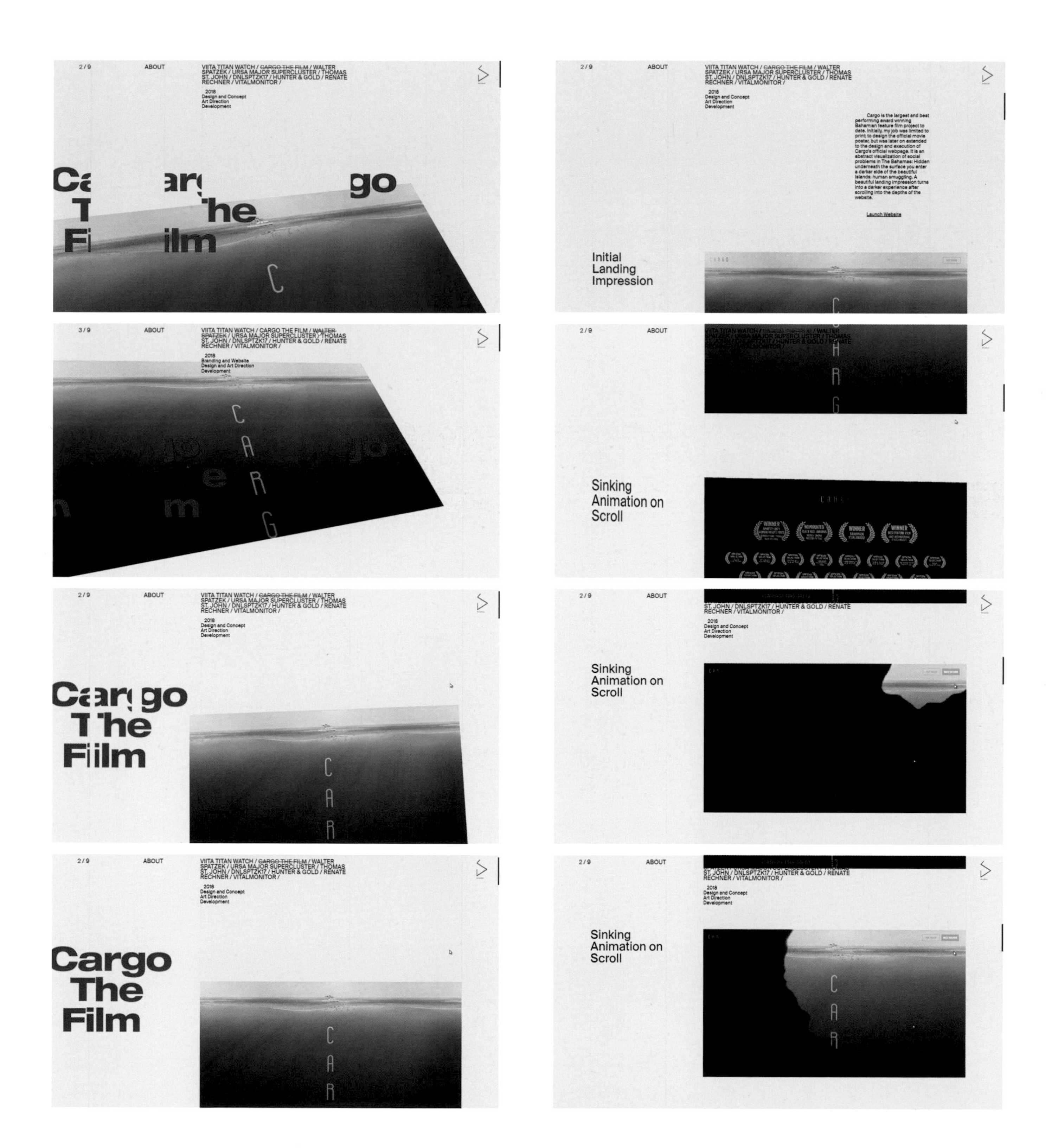

Behance

X / 9

~~ABOUT~~

VIITA TITAN WATCH / CARGO THE FILM / WALTER SPATZEK / URSA MAJOR SUPERCLUSTER / THOMAS ST. JOHN / DNLSPTZK17 / HUNTER & GOLD / RENATE RECHNER / VITALMONITOR /

ALREADY CONVINCED? 👇

office🙊danielspatzek.com

No? Ok, I see. Check out my awards & recognitions:

CSS Design Awards

1x Interactive Designer of The Year (2017)
1x Site of The Year Kudos Award (2017)
9x Site of the Day
1x Part of the judging panel - Senior Judge

MoVida Website

BY Zoë Barber

Scattered across Melbourne's CBD, the restaurants and bars of MoVida are as synonymous with the city as the laneways they inhabit. Within these lanes, local and international graffiti artists contribute to an ever-changing backdrop, while MoVida's iconic signage—vertically typeset and vividly red—remains a constant.

The typographic design of the MoVida's website pays homage to these two elements—kaleidoscopic-found lettering and bold vertical type. The animation treatments throughout take the signage reference further with individual glyphs appearing as if switched on, illuminating texts and images as the user browses down each page.

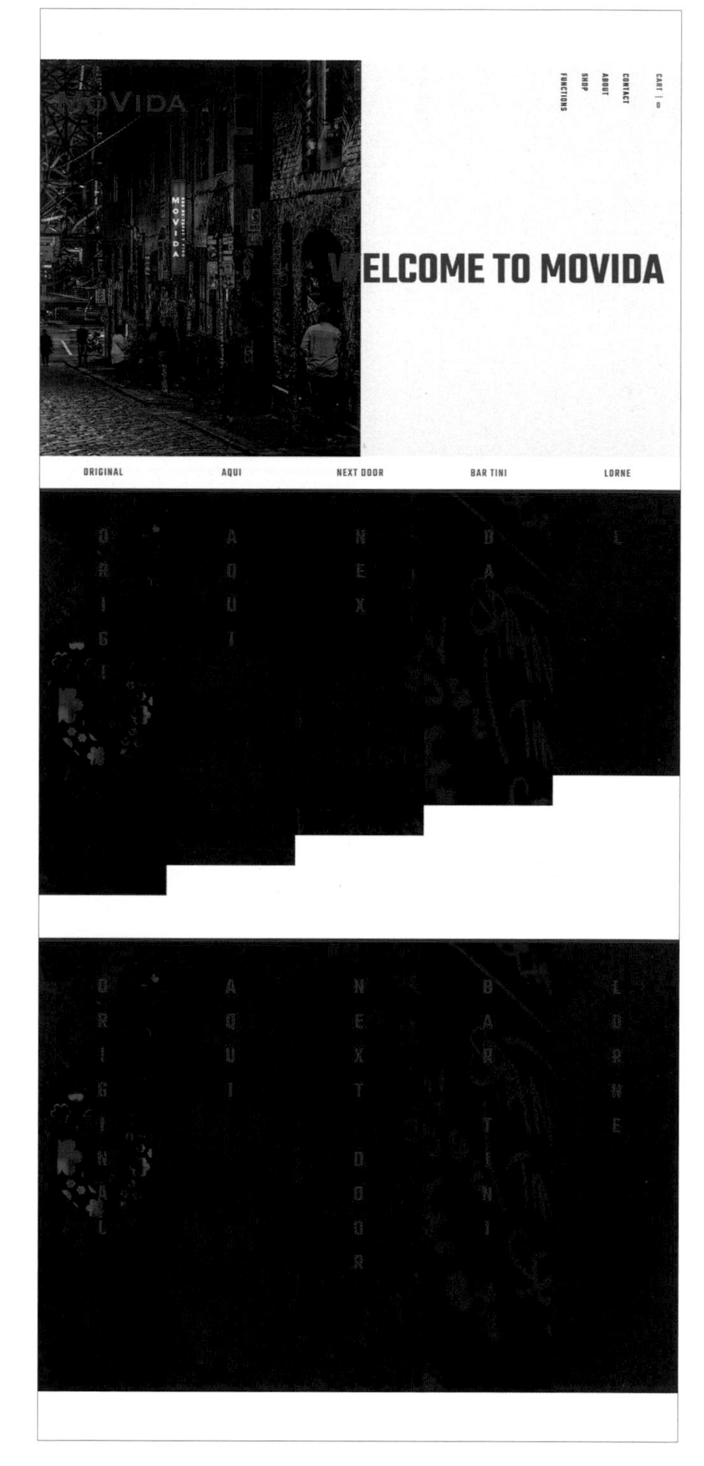

DEVELOPMENT **Ben Lovecek** / PHOTOGRAPHY **Alan Benson**

Official

Stanley Kubrick Biography Website

BY Tubik Studio

On the July 26th 2018, the date that marked the 90th anniversary of Stanley Kubrick's birthday, Tubik Studio presented all the fans of this famous film director, screenwriter, and producer with an elegant biography website devoted to his path of glory life and creative heritage in the world cinematography. Tubik Studio utilised the bold typography and contrast colours within a limited palette, high readability, and variety of visual effects.

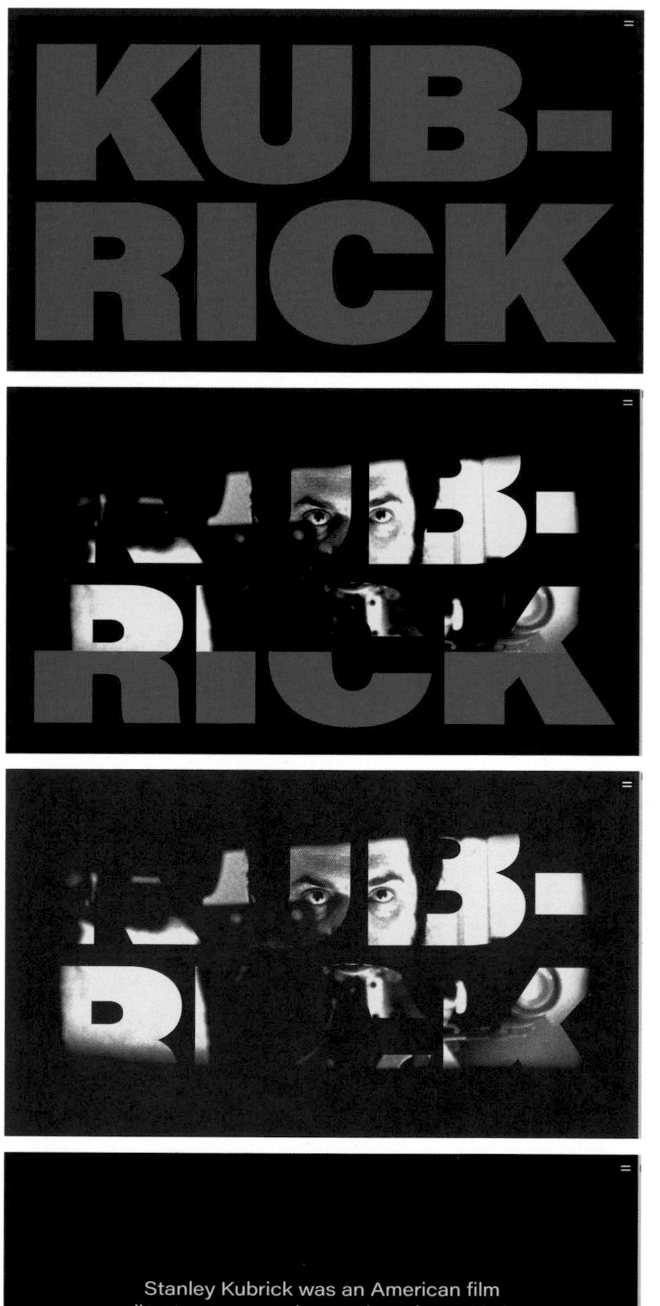

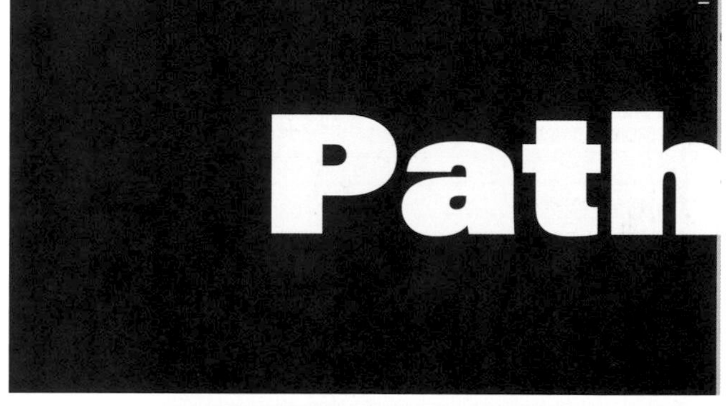

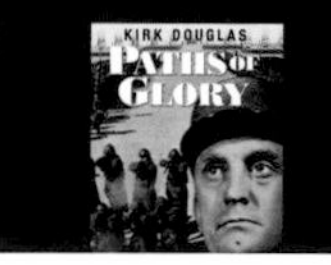

The tag 'Stanley Hubris'

Lolita's age was increased from 12 to 14

Sue Lyon as Lolita, 1960
© Bert Stern

Lolita

Xtian Miller's Official Website

BY Xtian Miller

Through unorthodox use of type, scale, and animation, Xtian Miller invoked a refreshing website user experience and angular style with American Text and Suisse Int'l SemiBold. Xtian incorporated the animations in both the type and background to create the feeling of disruption and movement, with infinite scroll to connect the end with the beginning.

Official

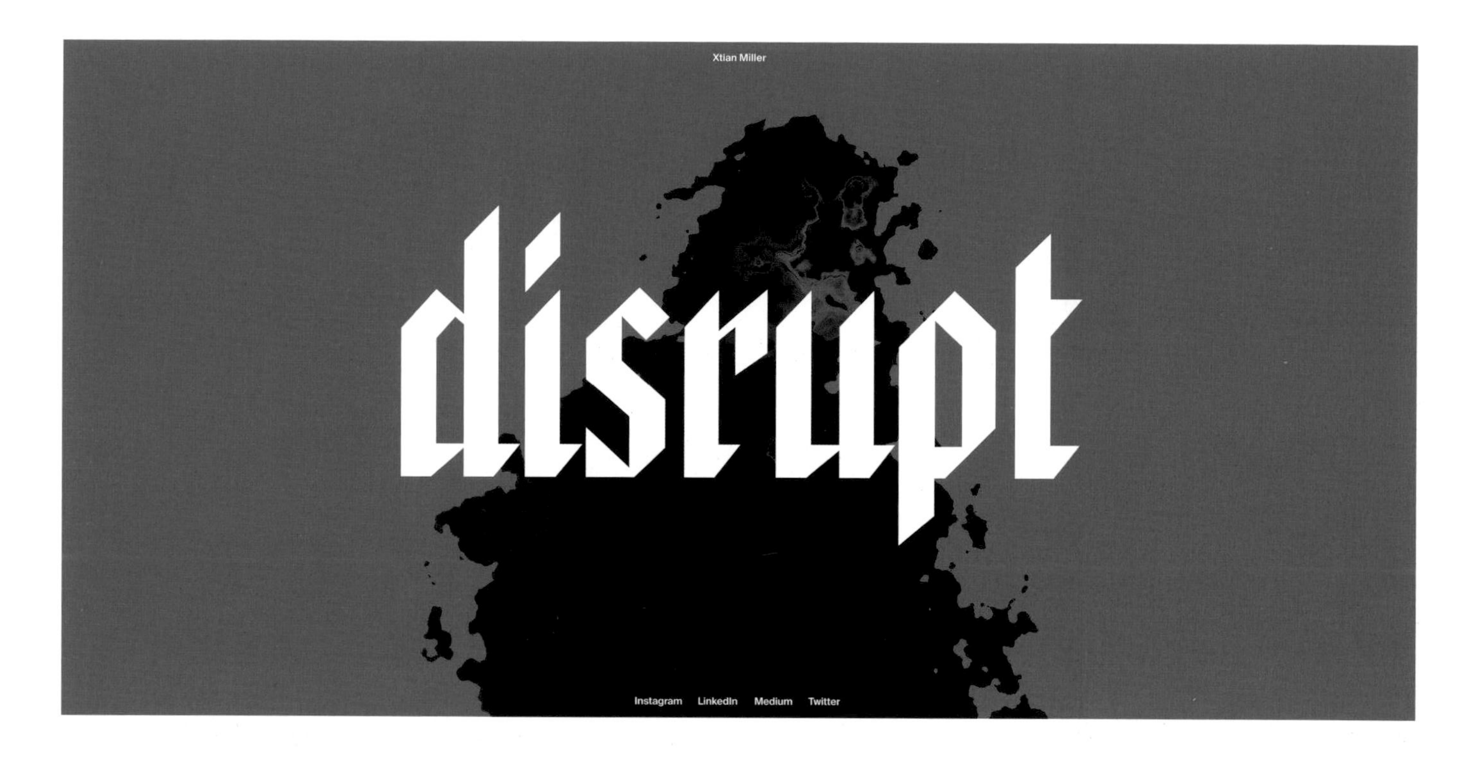

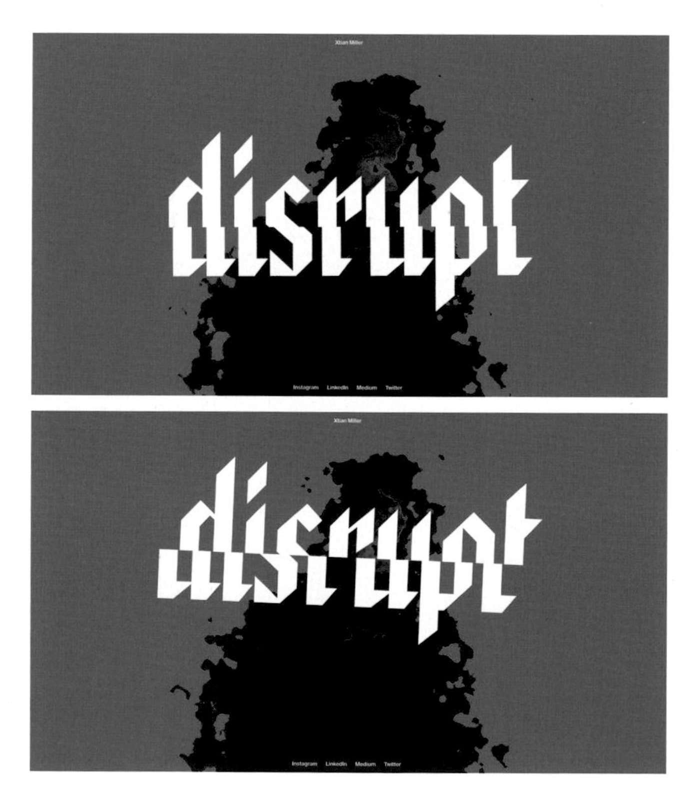

Inculerate

BY Red Collar

To reflect the image of Inculerate Inc. engaged in angel investments, Red Collar came up with the key visual of folding a paper plane to highlight the ground rules of Inculerate. Red Collar also chose the modern, neat, and easily readable Teko of Google Fonts to match but not to overshadow the key visual.

CLIENT **Inculerate Inc.**

DEMO Festival

BY Studio Dumbar

DEMO founded by Studio Dumbar (Part of Dept) and Exterion Media Netherlands is a festival celebrating the finest motions from the finest studios, designers, upcoming talents, and art academies from all around the world, showcasing works for 24 hours on all 80 digital screens located in Amsterdam Central Train Station.

The DEMO's identity utilises bold typography in a very fluid way—always moving and transforming. The logotype continuously transforms on screen and the movement is always contained within a set frame, referring to screen parameters.

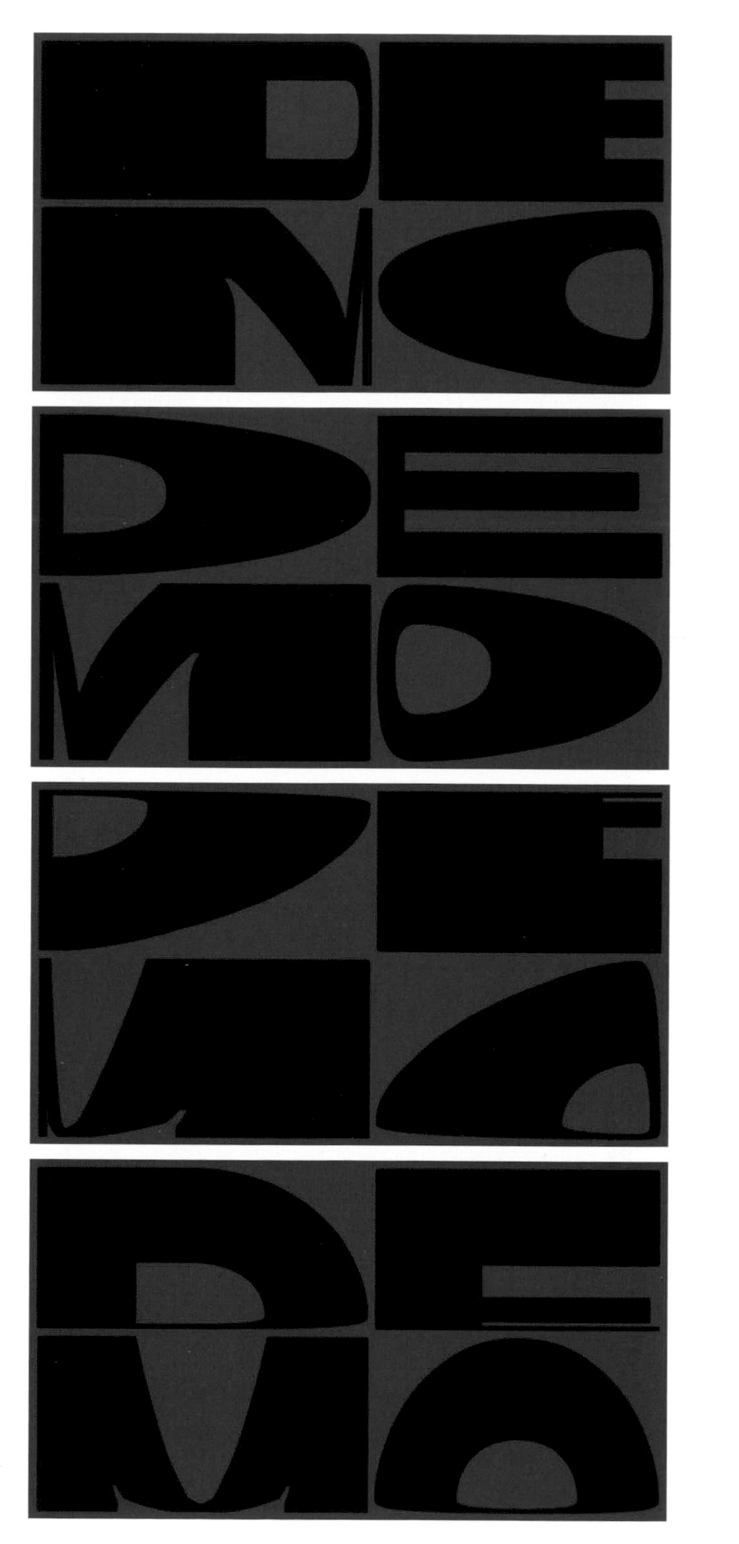

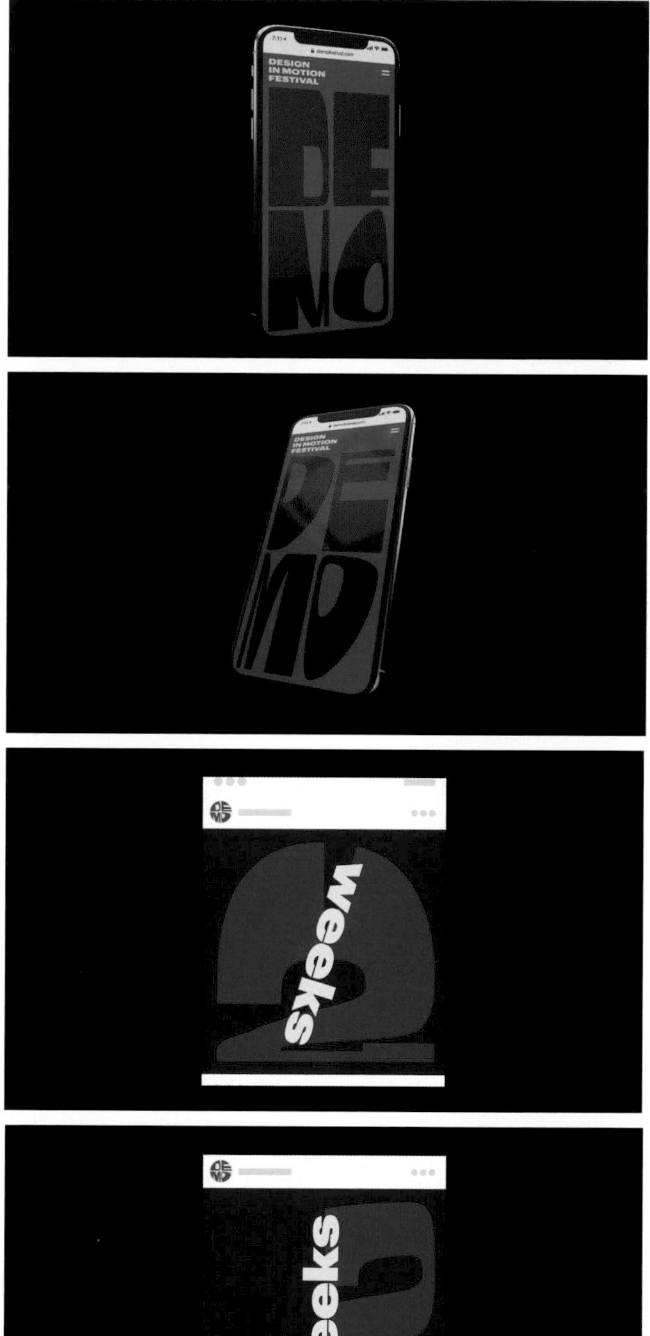

WEBPAGE

Official

12345678
9 ABCDEFGH
IJKLMNOPQ
RSTUVWXYZ

OPEN

CALL
CALL
CALL
CALL

OPEN

Contemple

BY Contemple

This project is regarded as the official website of Contemple. Contemple came up with the main design idea of using the downtime of the website in a more creative way. Several loading scenes and transition effects consist of the white intensive kinetic typography that can draw the viewers' attention.

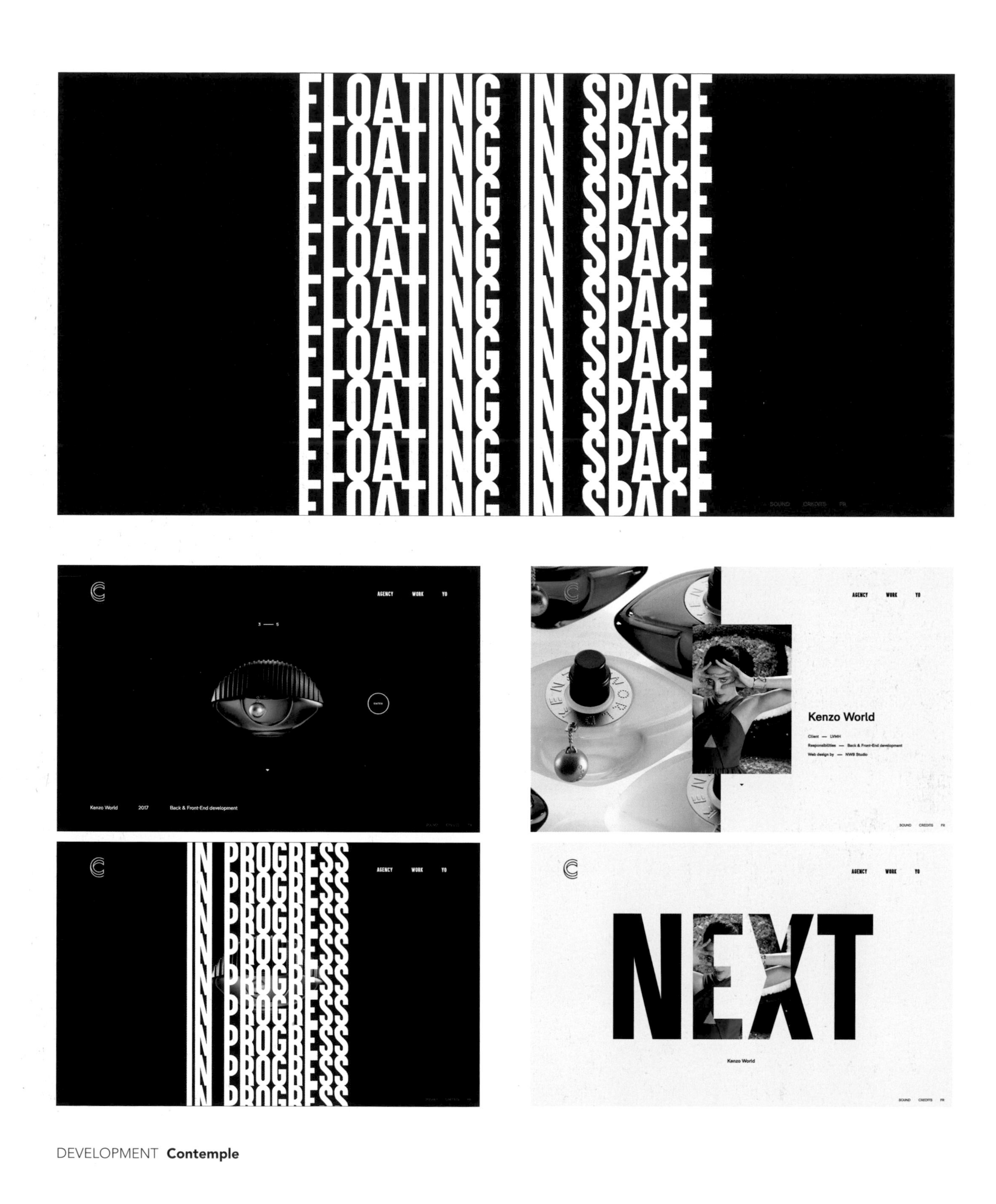

DEVELOPMENT **Contemple**

WEBPAGE

Official

AGENCY WORK YO

Contemple.
We are a creative agency based between
Paris and Montréal.
We provide a wide array of creative and
strategic solutions for brands.
We love music, design and internet.

SOUND CREDITS FR

Romain Granai's Personal Website

BY Romain Granai

Romain Granai did not design his own personal website by traditional techniques. He designed the website in the browser by experimenting with Cascading Style Sheets (CSS). And Romain found that a few lines of code can achieve a very interesting effect on the website. When a viewer scrolls down the webpage and moves the mouse, the typography's position, angle, and shape will change simultaneously. Meanwhile, the switch button on the website's top can transform the website between positive and negative colours.

Official

Romain Granai
design
Belgium.
currently living and
orking in London/UK.

Tim Roussilhe Portfolio

BY Timothée Roussilhe

This website is Timothée Roussilhe's portfolio. Timothée tried to develop a very minimal and elegant website with a modest design and type embellishments that help break the monotony of the simplistic layout without distracting from the featured works. The progression of this website is linear. As users scroll, they can browse and interact with different sections. In addition to this minimal structure, Timothée spent lots of time optimising the details and micro-interactions to delight the users.

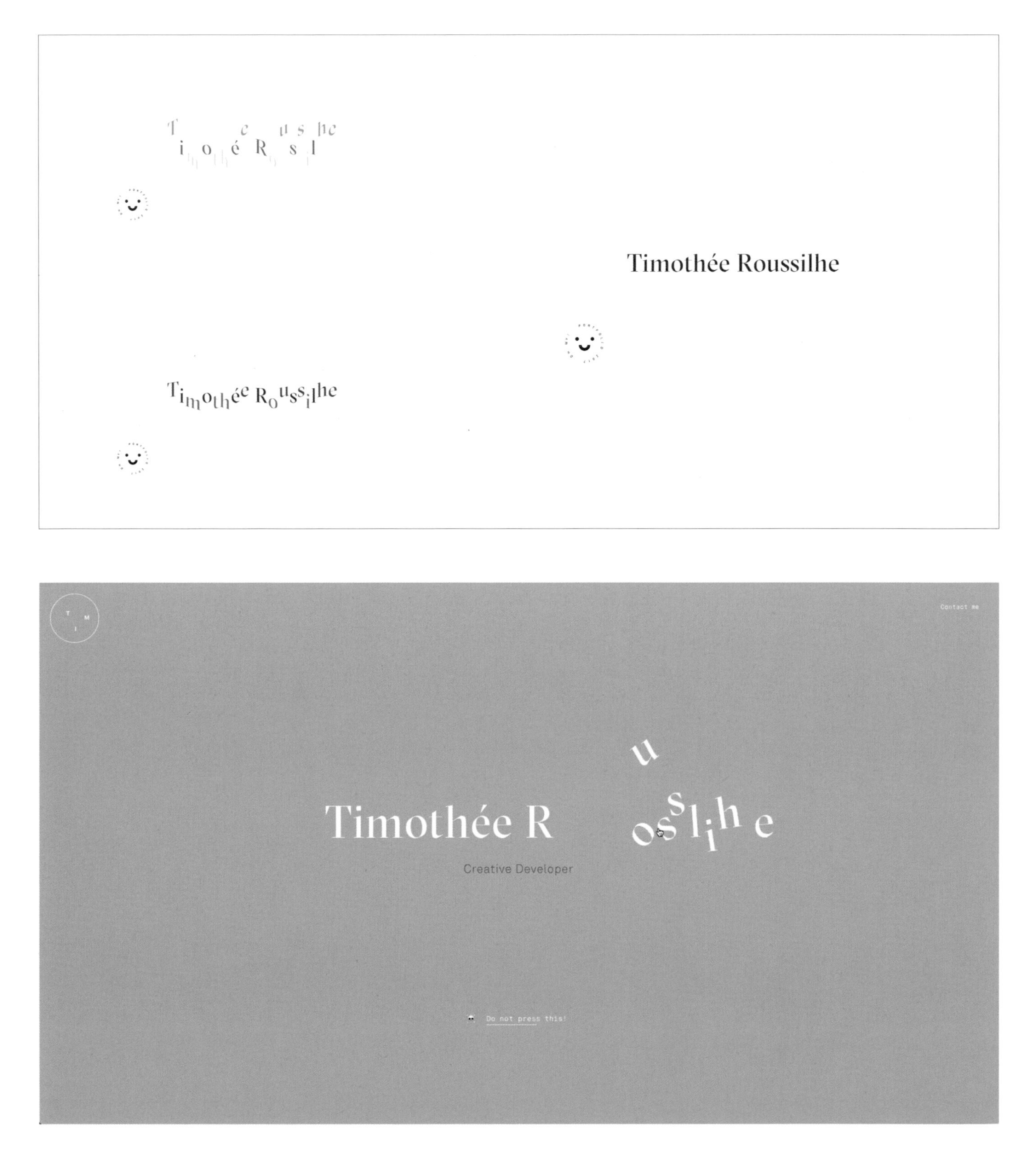

DEVELOPMENT **Timothée Roussilhe**

Official

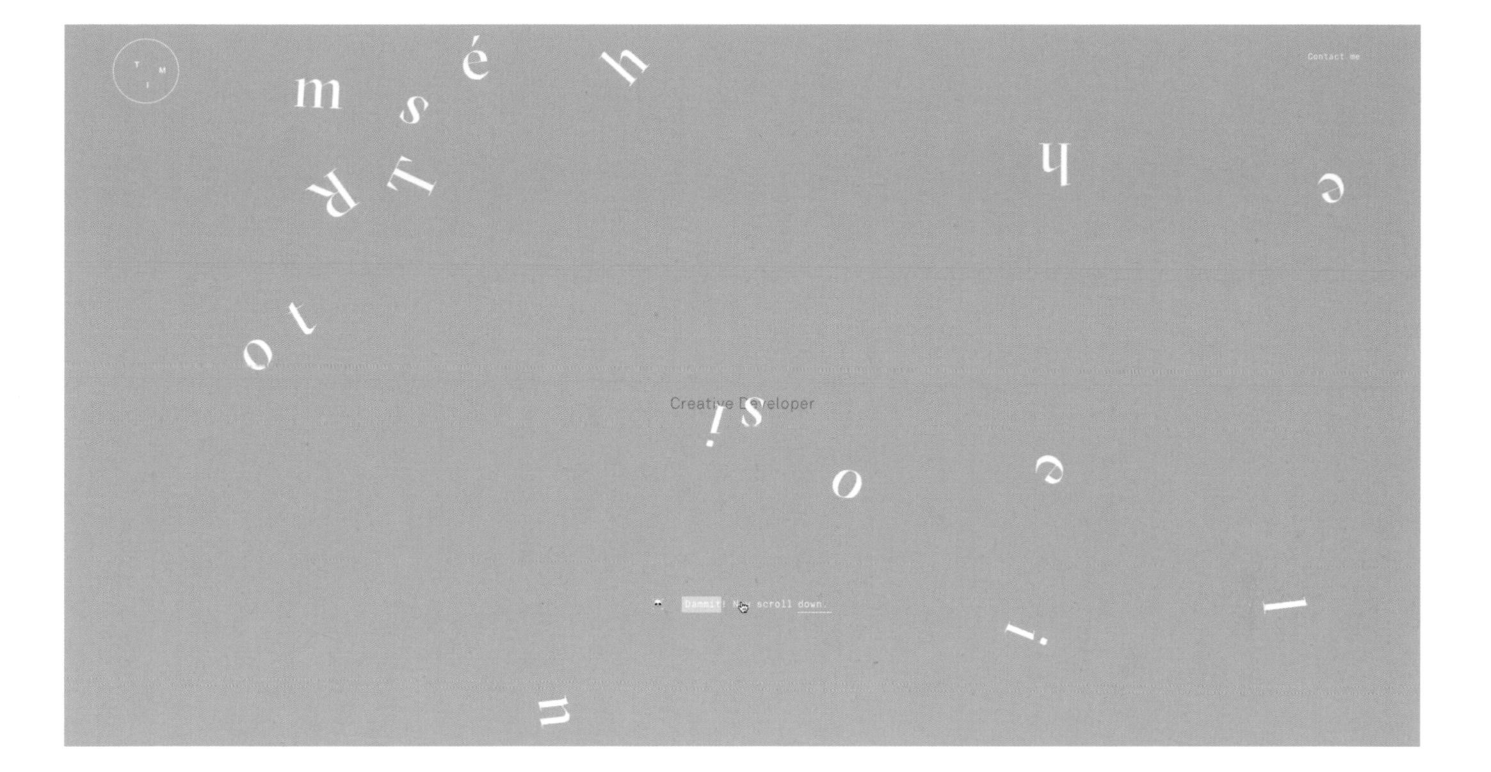
Contact me
Creative Developer
Dammit! Now scroll down

Vincent Tavano
LOEWE
JACK DAVISON CREATIVE DIRECTION FOR THE LONDON-BASED MOST TALENTED PHOTOGRAPHER.
Purpose & Form

BLK OUT

BY BLK OUT

The design idea of this official website is to create and display the identity and mindset of BLK OUT. This website is a perfect mix of creativity and technics. BLK OUT has created glitch effects, 3D movements, and immersive transitions. And this website has to reveal the unboring spirit of BLK OUT.

Official

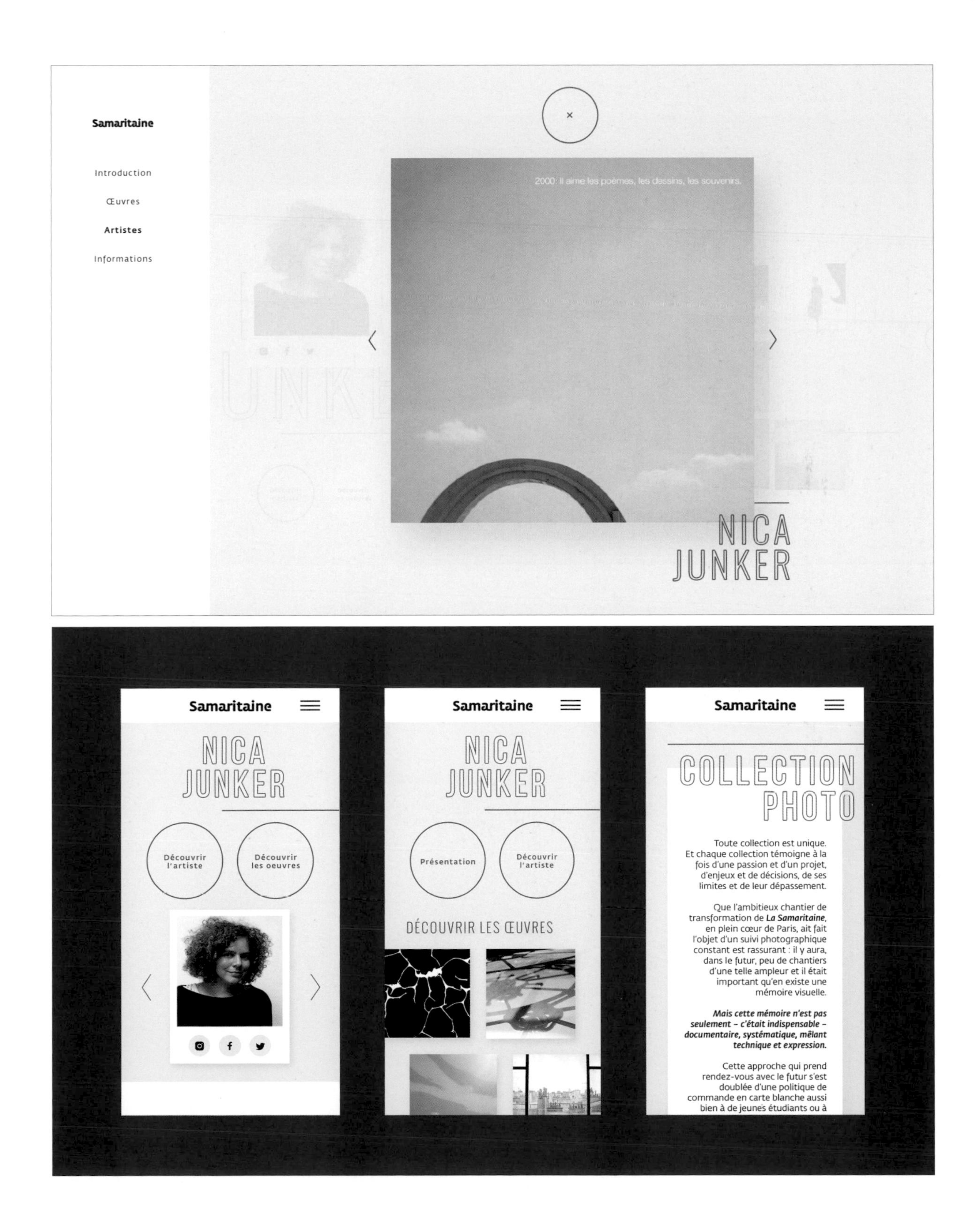
Samaritaine
Introduction
Œuvres
Artistes
Informations
2000: Il aime les poèmes, les dessins, les souvenirs.
NICA
JUNKER
Samaritaine
NICA
JUNKER
Découvrir l'artiste
Découvrir les oeuvres
Samaritaine
NICA
JUNKER
Présentation
Découvrir l'artiste
DÉCOUVRIR LES ŒUVRES
Samaritaine
COLLECTION
PHOTO
Toute collection est unique. Et chaque collection témoigne à la fois d'une passion et d'un projet, d'enjeux et de décisions, de ses limites et de leur dépassement.
Que l'ambitieux chantier de transformation de La Samaritaine, en plein cœur de Paris, ait fait l'objet d'un suivi photographique constant est rassurant : il y aura, dans le futur, peu de chantiers d'une telle ampleur et il était important qu'en existe une mémoire visuelle.
Mais cette mémoire n'est pas seulement – c'était indispensable – documentaire, systématique, mêlant technique et expression.
Cette approche qui prend rendez-vous avec le futur s'est doublée d'une politique de commande en carte blanche aussi bien à de jeunes étudiants ou à

Safari Riot: Noise Website

BY Safari Riot

Safari Riot intended to create something playful and memorable on their official website. The website's design was meant to convey Safari Riot's creatively innovative aesthetic while still being functional for people interested in diving deeper into what they do and how to work with them. The key part of the design process for Safari Riot was to consider usability first, and then immediately reconsider whether the usability assumption was based on the common practice.

CREATIVE DIRECTION **Grayson Sanders, Jerry Yeh** / UX/UI DESIGN **Jeany Ngo** / DEVELOPMENT **Eric Van Holtz**

WEBPAGE

Official

UNDER ARMOUR
ALIEN:COVENANT
Sound + Music for the Wild
BLADE RUNNER
Trailer
"Announcement", "Trailer #2", "Int. Trailer"
– Pusher/Wildcard |
12.19.2016
BROWSE
ABOUT
CONTACT
#DAILYBOUNCE
:XP
NEW YORK CITY
SAFE&SOUND
LINA
BROWSE
ABOUT
MUSIC
SOUND
CODE
STUDIOS
CONTACT
#DAILYBOUNCE
:XP

souffl.com

BY Souffl

This is the portfolio website of Souffl. They present their design works, offers, and value proposition. When designing this website, they used only one font—Souffl Ultra Condensed which was designed to be responsive. Such a font became the structure of the website graphic design.

DESIGN **Nicolas Baumgartner, Arnaud Carrette, Fabien Fumeron**

Official

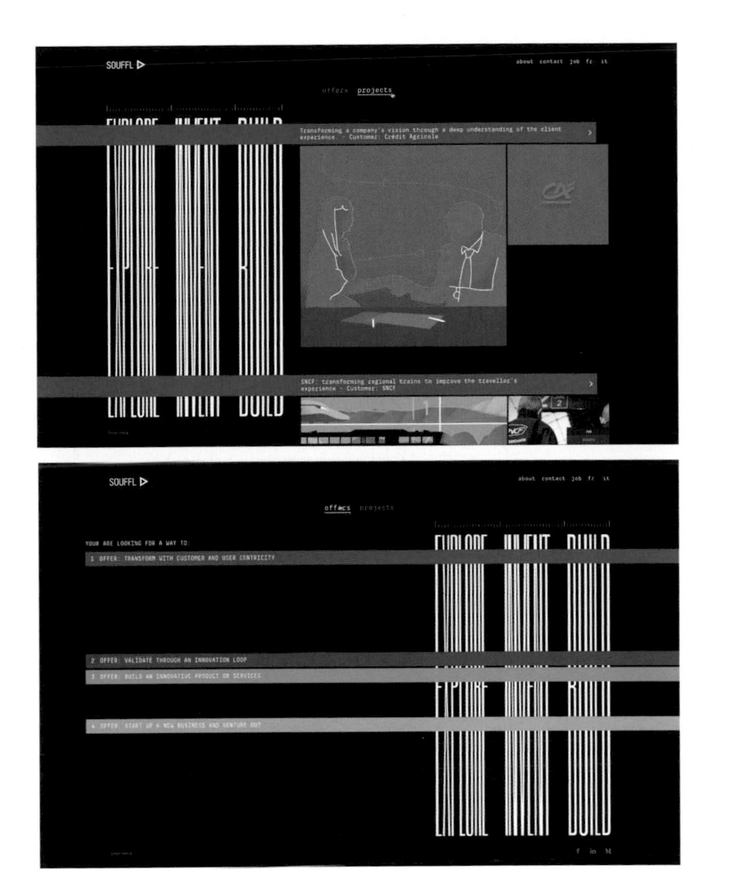

SOUFFL
about contact job fr it
offers projects
YOUR ARE LOOKING FOR A WAY TO:
1 OFFER: TRANSFORM WITH CUSTOMER AND USER CENTRICITY
2 OFFER: VALIDATE THROUGH AN INNOVATION LOOP
3 OFFER: BUILD AN INNOVATIVE PRODUCT OR SERVICES
4 OFFER: START UP A NEW BUSINESS AND VENTURE OUT

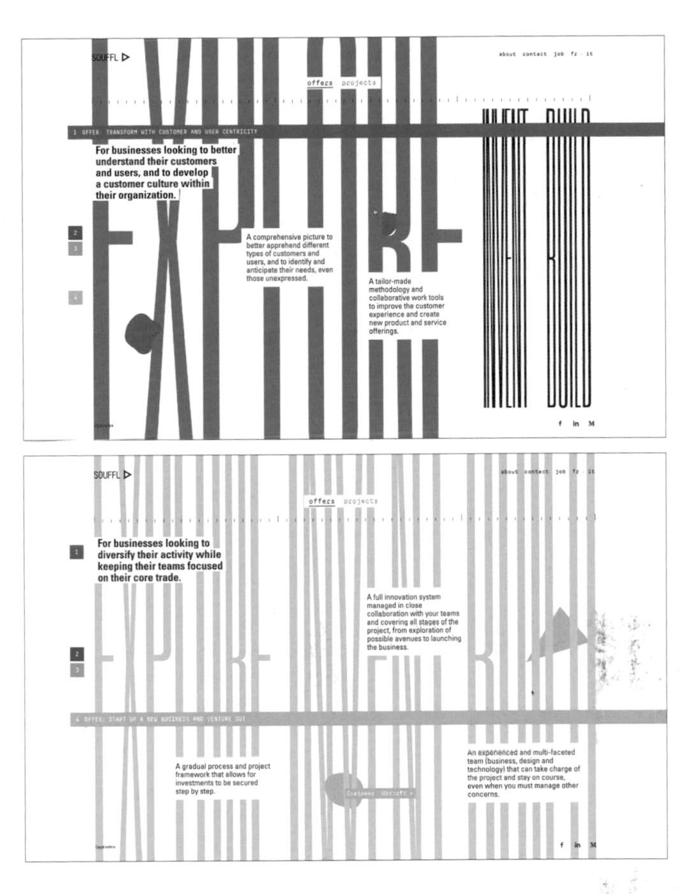

SOUFFL
about contact job fr it
offers projects
1 OFFER: TRANSFORM WITH CUSTOMER AND USER CENTRICITY
For businesses looking to better understand their customers and users, and to develop a customer culture within their organization.
A comprehensive picture to better apprehend different types of customers and users, and to identify and anticipate their needs, even those unexpressed.
A tailor-made methodology and collaborative work tools to improve the customer experience and create new product and service offerings.
For businesses looking to diversify their activity while keeping their teams focused on their core trade.
A full innovation system managed in close collaboration with your teams and covering all stages of the project, from exploration of possible avenues to launching the business.
A gradual process and project framework that allows for investments to be secured step by step.
An experienced and multi-faceted team (business, design and technology) that can take charge of the project and stay on course, even when you must manage other concerns.

SOUFFL
offers projects about contact job fr · it
April 2019 —
Inventing a New Business
Customer: Paris-Turf
From Specialized Press to the Internet of Things.
January 2016: Souffl teamed up with the general management of the Paris-Turf Group, leader of the equestrian press, to launch a bold new project: reinventing the future of their business. December 2017: less than two years later, the Paris-Turf Group announced the launch of Exalt training, the first performance tracking solution developed with racehorse trainers in mind. Exalt training is the result of detailed scientific research, precise engineering and demanding design, carried out by Souffl over the course of two years (cf. Case study:the development of Exalt training). It also represents the creation of a new business, the culmination of a fantastic innovation journey started from scratch and undertaken hand in hand with the teams at Paris-Turf. Here is the story of this entrepreneurial adventure.
Speciality
Diversification strategy
Stage gate process
Identifying and enabling opportunities
Intrapreneurship guidance
Business model design
Designing and validating the offer
Brand design

BingePlease X Netflix

BY Camille Frairrot

BingePlease is an interactive web game. This game consists of a series of ten visual riddles. The players need to find out the names of those Netflix TV shows according to the specific illustrations as quick as possible. After obtaining their final score, the web game leads the players to discover their "binge profiles" and TV shows' recommendation.

COOPERATION **Adrien Vanderpoote, Antonin Rivère, Victorie Douy**

Behance

BINGE PLEASE
BINGE PLEASE
BINGE
BINGE PLEASE
BINGE
BINGE
BINGE
BINGE
BINGE-PLEASE.COM
BINGE
BINGE PLEASE
BINGE
BINGE
BINGE
BINGE
BINGE
e please
Binge ple
BINGE-PLEASE.COM
NETFLIX

Gyro Brand Website

BY Bekk

Bekk was asked for a website project for Gyro. And quickly the website project became a development for a new visual identity. Bekk focused on the people who work in Gyro, their workflow, and their goals and values—"We accelerate people, cultures, and brands." Getting inspired by the mechanism of gyroscope, Bekk made a playful intro with "Gyro" on the new website. The four letters "Gyro" will be disassembled automatically and follow the movement of the mouse to perform a free rotation in all three axes as if an invisible rotor maintains the gyroscope's spin axis direction.

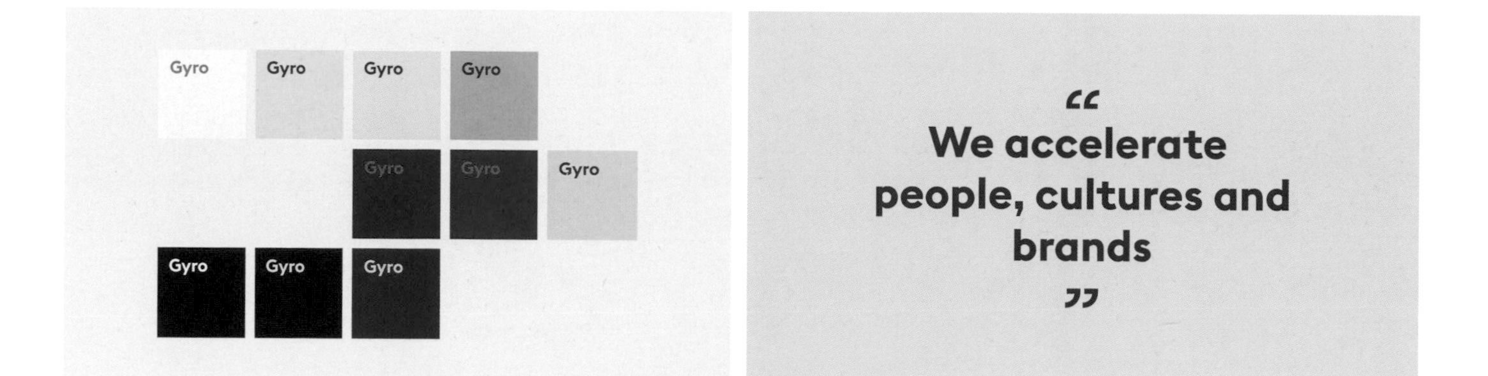

DESIGN **Hans Christian Øren** / MANAGEMENT **Nina Inberg**
DEVELOPMENT **Knut Erik Langdahl** / CLIENT **Marius O.W. Kveim, Fredrik Hasselheid**

Behance
Official

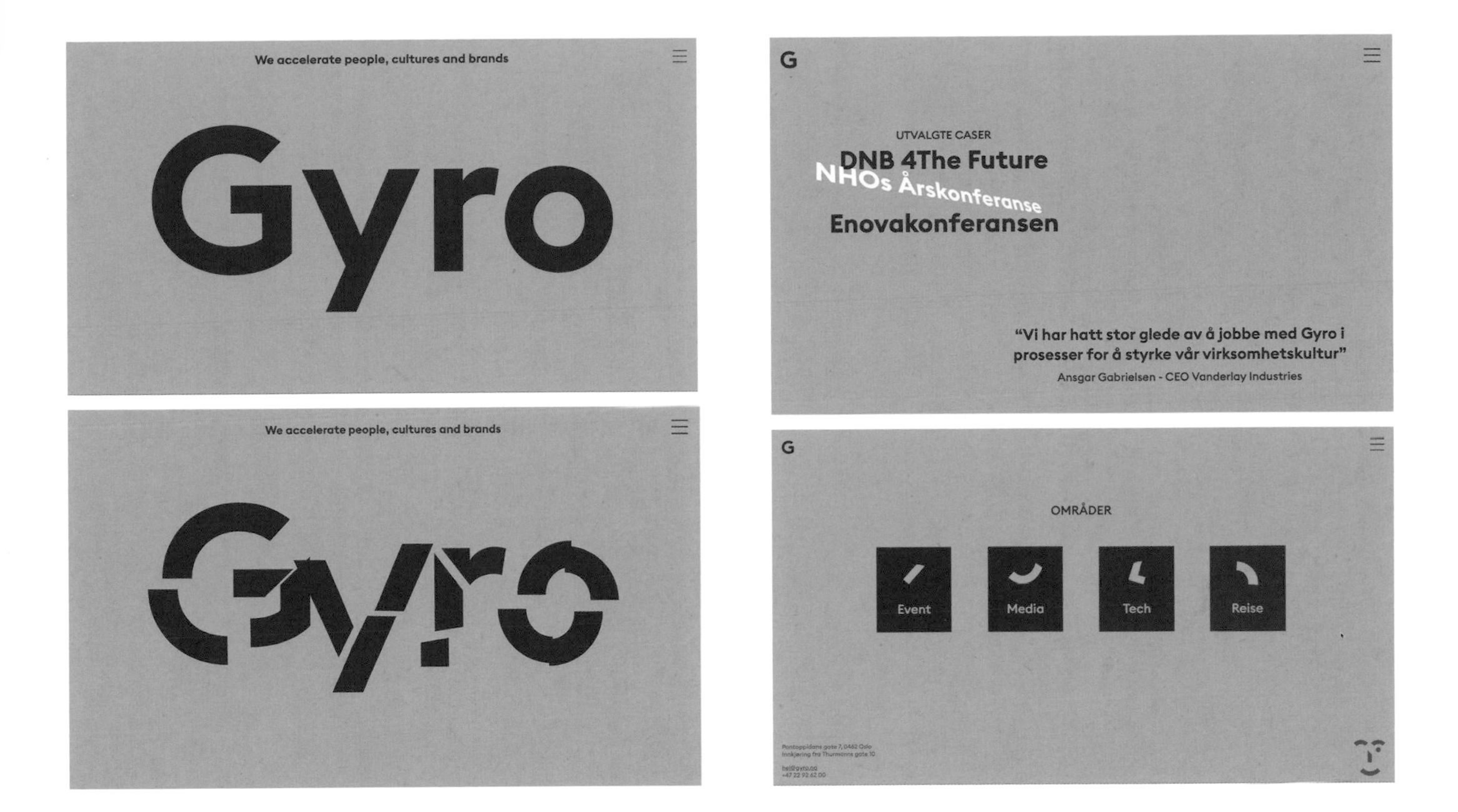
We accelerate people, cultures and brands
Gyro
G
UTVALGTE CASER
DNB 4The Future
NHOs Årskonferanse
Enovakonferansen
"Vi har hatt stor glede av å jobbe med Gyro i prosesser for å styrke vår virksomhetskultur"
Ansgar Gabrielsen - CEO Vanderlay Industries
We accelerate people, cultures and brands
Gyro
G
OMRÅDER
Event
Media
Tech
Reise

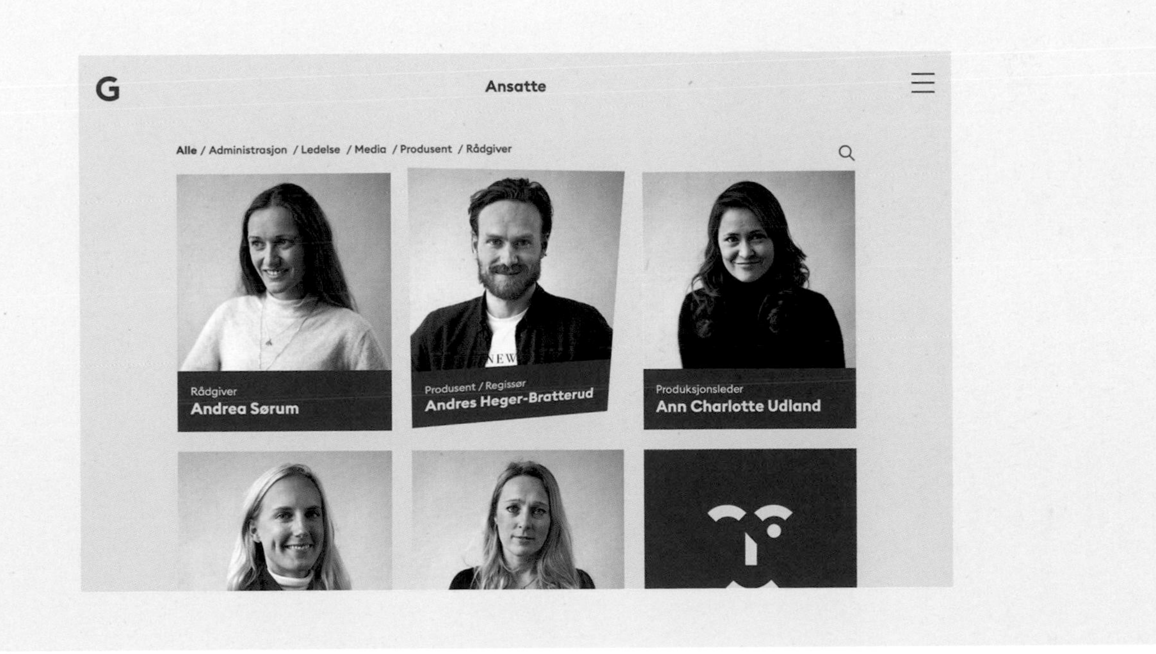
G
Ansatte
Alle / Administrasjon / Ledelse / Media / Produsent / Rådgiver
Rådgiver
Andrea Sørum
Produsent / Regissør
Andres Heger-Bratterud
Produksjonsleder
Ann Charlotte Udland

Particles Typeface

BY CLEVER°FRANKE

The weather has always been a very important inspiration for the works of CLEVER°FRANKE. The degree sign in their name—a reference to the astronomer Anders Celsius (°C) and the physicist Daniel Fahrenheit (°F)—serves as the basis for the design of their visual identity. This typography was primarily inspired by the wind. It is designed to create an effect of flowing and guide the visitors through their website.

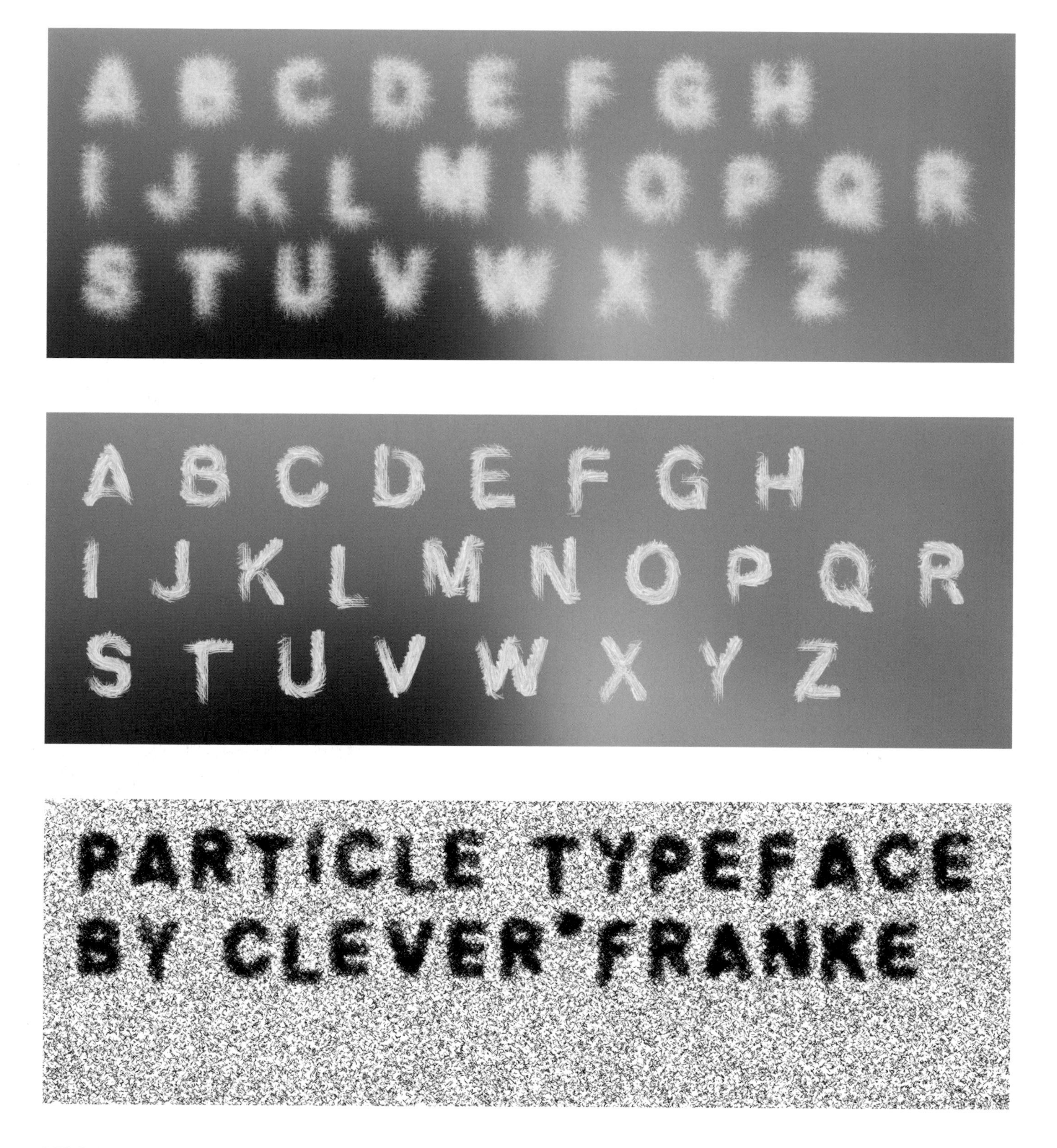

DESIGN **Gert Franke, Thomas Clever** / DEVELOPMENT **Davey van der Woert**

WEBPAGE

Official

C°F
HOME
WORK
APPROACH
ABOUT
JOBS AND TEAM
CONTACT
WE CREATE DATA DRIVEN TOOLS AND EXPERIENCES
WATCH OUR STORY
Google
here
WIRED
ELSEVIER
WARNER MUSIC GROUP
C°F
Illustrations
Qbit Identity
Chicago's Mobility
Smart Watch Interface
Internet of Elephants
Consumer Barometer
Migration data
Trendviz
Living Cell
Iris.ai
My Research Dashboard
APPROACH
ABOUT
JOBS AND TEAM
CONTACT
WIRED
VIEW CASE STUDY

JA Studio

BY Bahaa Samir

Jafen Art (JA) is a design studio driven by the creation of unique, distinctive, and memorable brand communication. The designers came up with a highly creative design that features the nice typography merging between modern and contemporary art. The designers also have used the letters "JAFEN" to make an interactive website introduction.

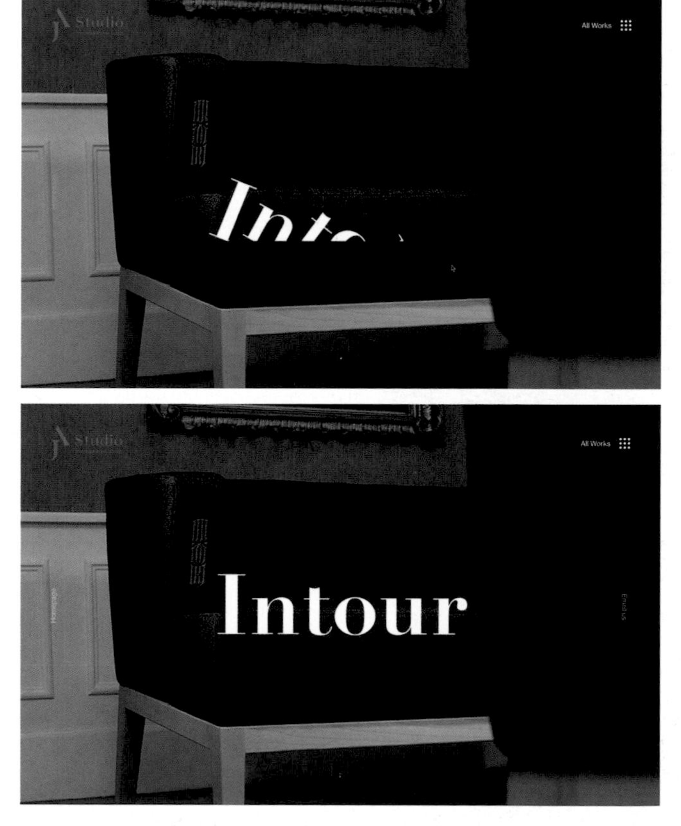

DESIGN **Hisham Mohamed** / DEVELOPMENT **Bahaa Samir** / CLIENT **Jafen Art Studio**

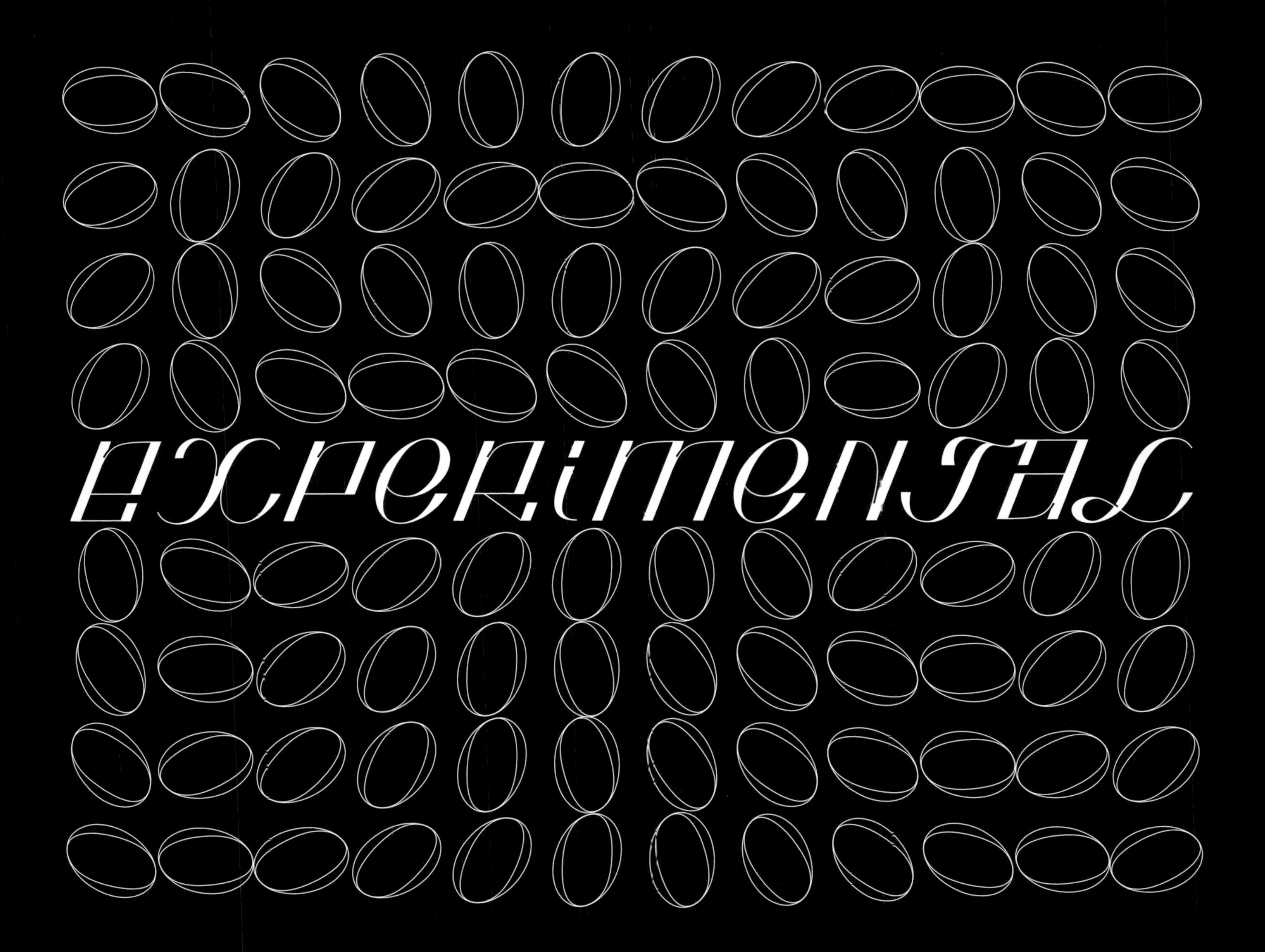
Experimental

Interview with
DUNCAN BRAZZIL

Duncan Brazzil (Design By Duncan) is an American/British graphic designer specialising in motion, typography, digital, print, and brand identity residing in Edinburgh, Scotland. Currently residing in Portland, Oregon, he is exploring the subject of kinetic typography with highlighted projects "36 Days of Type" and "Kinetic Typography Experiments."

How, when, and why did you start working in the type design field? Can you share some design experiences with us?

I started working on kinetic typography when I first picked up Adobe After Effects and experimented on it. From the beginning of pursuing Graphic Design, I have always been interested in type as a key visual element aesthetically and felt amazed by the motion graphics. Since I have taken the kinetic typography on, I definitely feel more of a connection with typography than ever before, though I still would not refer to myself as a motion designer.

The first project I took on with kinetic typography was the Adobe's 36 Days of Type Challenge in 2018. Ever since then, I have been experimenting almost on a daily basis, constantly trying to improve and make it become more technical. At the same time, I can still get a better skill and that becomes the reason why I am having so much fun in forming my own style.

The typefaces you designed, such as "36 Days of Type" and "Kinetic Typography Experiments," have amazing variable effects. What motivated and inspired you to create those typefaces? In terms of the design techniques and aesthetics, how do variable typefaces differ from static ones?

Experimentation purely allowed me to create those effects—I would not even be able to tell anyone how I achieved some of them in the end. From the beginning of creation, I have realised that some of the knowledge I have learned from the relevant tutorials on YouTube has been chopped up in practical operation and I need to put my own slants on a lot of effects. It does not mean that I ripped the ideas. Instead, I would apply what I had learned in an abstract way and dig deeper until I could end up with something delightful. Personally, variable typefaces represent more characteristics since they are animated.

In recent years many typography designers have spontaneously started various projects of experimental typefaces on Behance, Instagram, and so on. What do you think about this trend? And what is the main purpose and meaning behind such a trend in your opinion?

Social media is a great tool for young creatives to get noticed and increasingly becoming more of the fact. It is obvious that there are a lot of opportunities for sharing works on those social media platforms in 2019. And there is no doubt that the volume and propagation speed of those works are outrageous. I think that it is bad and unhealthy that creatives are just focusing on the thumb-ups or likes for their works on social media. I have definitely realised how important it is to break away from social media recently and get some inspirations from the outside world. It could be better than indulging in the daily posting and scrolling on Instagram.

In terms of the trend of kinetic typography, I have communicated with many others in this sort of emerging design community, such as DIA, Studio Dumbar, and designers like Xavier Monney (working with Kenzo) and Anthony Velen (working with Nike and Dropbox) and found that powerful kinetic typography has become a visual trend adapting to brands all over the world.

What is your workflow like? How did you prepare for your preliminary work?

I start off sketching ideas and storyboarding on paper then move to try and execute. That could save a lot of time. If sometimes I have an idea already formed in my head and want to get it out, I will try my luck to get it done if my technical ability can achieve.

What are the main challenges during the design of those typography experiments?

Time! That makes the series of "Kinetic Typography Experiments" valuable. Every single piece has been created in one hour or less usually. To me, it is almost like a log of all the effects I have explored and mastered in the relevant tutorials on YouTube from late nights.

In your opinion, what makes kinetic typography so impactful?

Words and movements within kinetic typography are so powerful. Kinetic typography will be even more impactful when its characteristics match with the product and brand.

With the development of mobile Internet, people do not just look for variable types but also responsive types so that they can be used on different digital screens. How do you look upon such a phenomenon?

All for the variable types! All the lovely type foundries have followed this trend. It is amazing to see this trend within the typography design impacting the motion graphics. And Colophon Foundry and Dalton Maag are very inspiring for us to learn variable types.

Could you share your upcoming projects with us?

I have just finished my "36 Days of Type 2019" as well as two big freelance projects—one of them was working with Design by Jake to create digital assets used throughout NYFW for the Society Model Management. Meanwhile, I have also had the opportunity to work with some incredible studios including Daughter Studio, Scottish Design Agency of the year, Cause & Effect, and Hype Type Studio in Los Angeles.

▾ **KINETIC TYPOGRAPHY EXPERIMENTS** (P158)

Motion Type Research

BY Alexander Slobzheninov

Motion Type Research is an experimental ongoing project, the goal of which is to explore the possibilities and beauty of kinetic typography out of any context. The animated quotes were inspired by Oblique Strategies, a wonderful method for promoting creativity. All the fonts in this project were designed by Alexander Slobzheninov, such as Agrandir, Grafier, Object Sans, and so on.

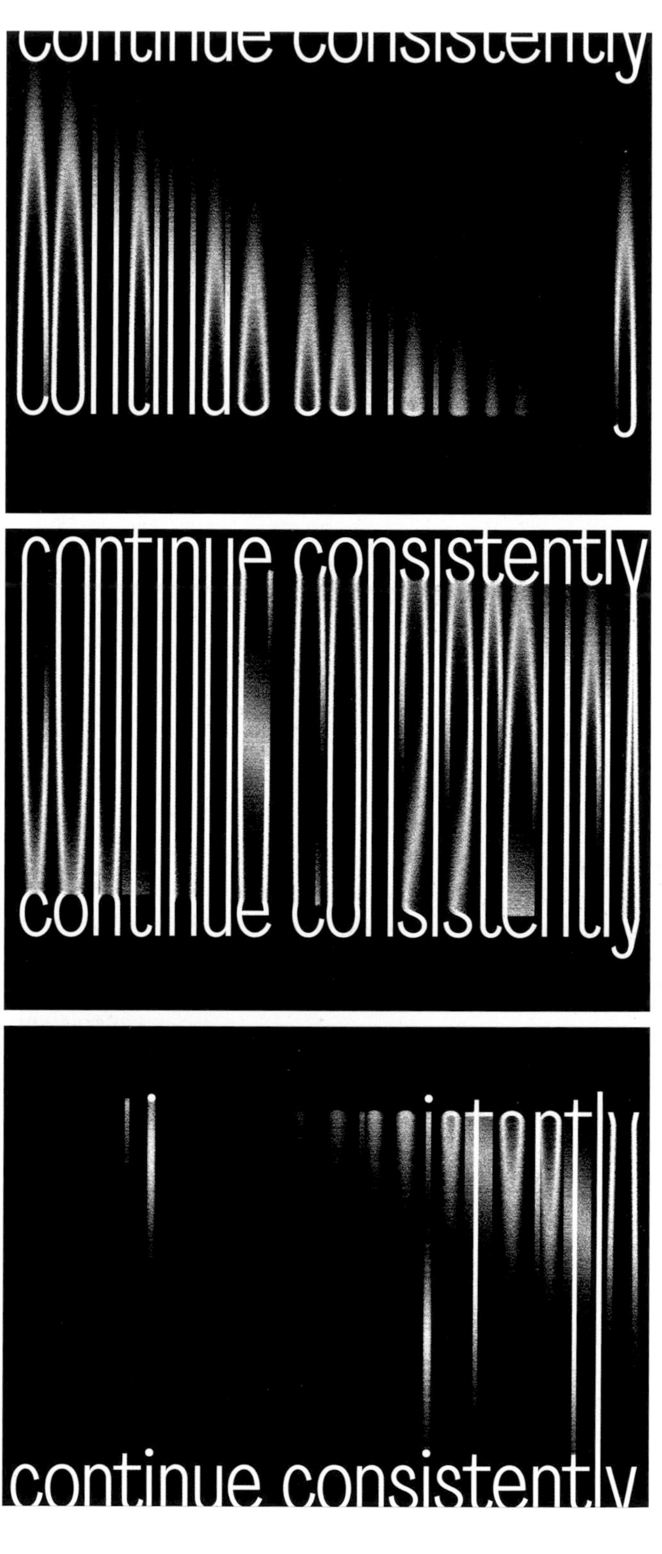

TYPOGRAPHY DESIGN **Alexander Slobzheninov**

Behance

THE SEARCH HAS BEG
BE FOUND
SOM
SOMETHING WI
ONCE THE SEARCH
SOMETHING WILL BE
BEGUN
ONCE THE SEARCH
SOMETHING
FOUND
SOMETHING HAS
SOMET

Motion Poster Series

BY Vince Hegedűs

Motion Poster Series is a collection of experimental posters within the field of kinetic typography and motion design. Every poster is based on a single word focusing on the dynamic effect of the motion referring to the meaning of the word itself. This series merges etymology and animation in a conceptual way.

Behance

AWA
REN
ESS
AWA
REN
VINCE HEGEDUS
TYPOGRAPHY EXPERIMENT
DATE
2018.06.25.
RAISE
AWARENESS
SWIPE
VINCE HEGEDUS
TYPOGRAPHY EXPERIMENT
DATE
2018.05.19.
BLANK POSTER
SWIPE
VINCE HEGEDUS
TYPOGRAPHY EXPERIMENT
DATE
2018.06.25.
RAISE
AWARENESS
VINCE HEGEDUS
TYPOGRAPHY EXPERIMENT
DATE
2018.05.19.
BLANK POSTER
SWIPE

Goertek Sonic Typeface

BY Daikoku Design Institute,
the Nippon Design Center, Inc.

Daikoku Design Institute designed a sound sensitive typeface for the Chinese acoustic components company Goertek and its R&D hub in Qingdao, China. This variable typeface synchronises sound waves and frequencies even in the digital signage. And Daikoku Design Institute developed the font in collaboration with Kontrapunkt in Denmark.

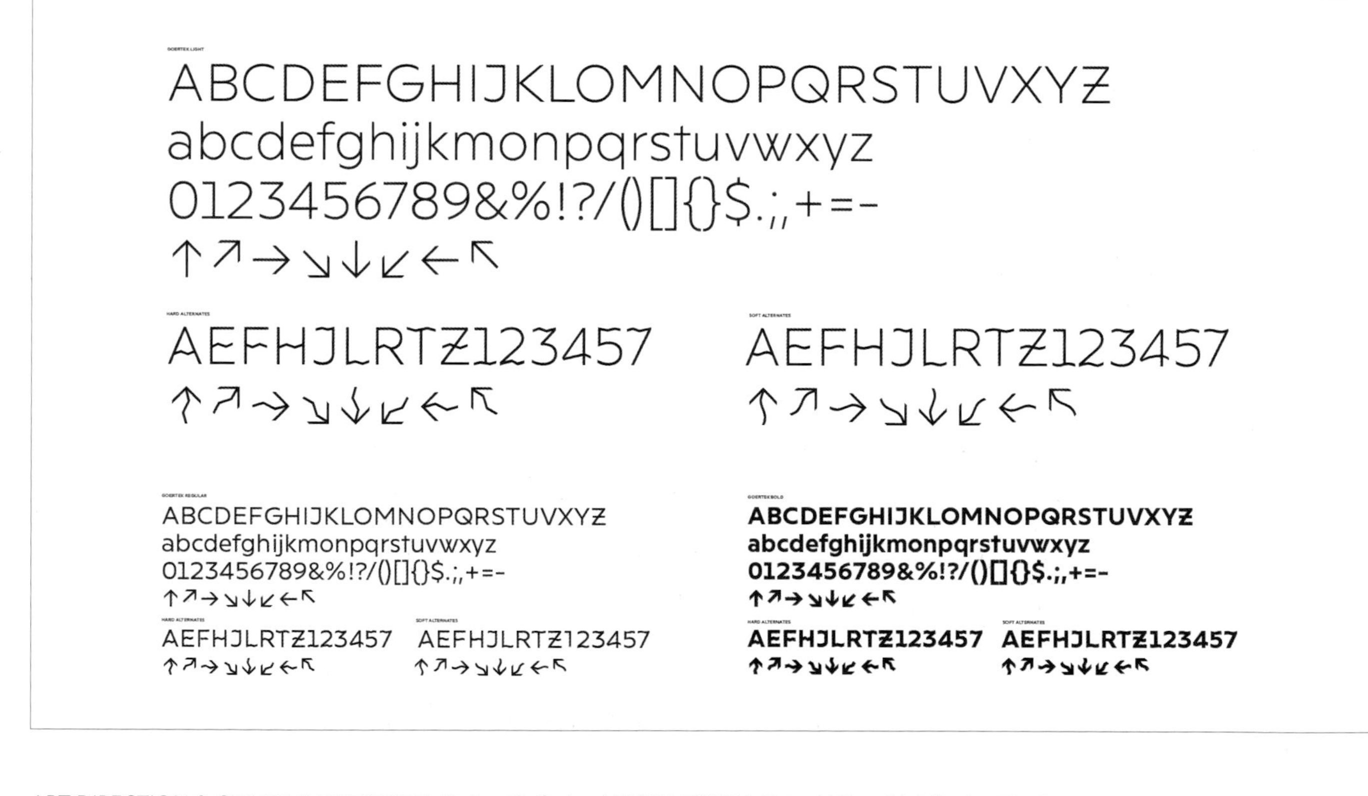

ART DIRECTION & CREATIVE DIRECTION **Daigo Daikoku** / PRODUCTION **Tatsuki Suzuki, Mizuho Morita**
TYPOGRAPHY DESIGN **Rasmus Michaëlis, Torsten Lindsø Andersen** / MANAGEMENT **Lars Larson** / CLIENT **Goertek**

Official

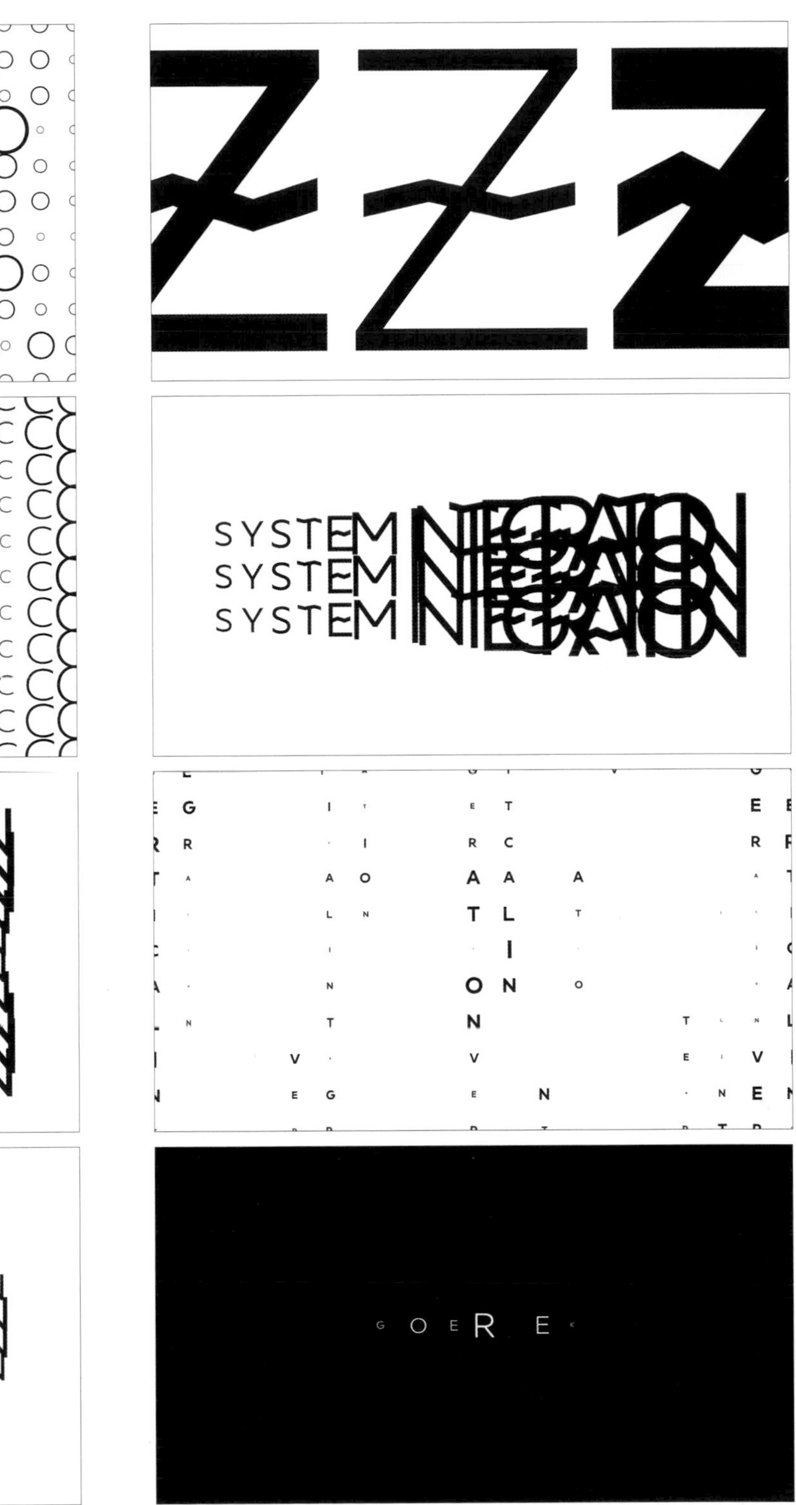

More–Typography

BY Ameen Shahid

Ameen Shahid got inspired by a book called *The Subtle Art of Not Giving a F*ck* and extracted the element "more" from the sentence—"The world is constantly telling you that the path to a better life is more, more, more..." As an audio-driven motion design, Ameen decided to make a snappy animation by stretching and distorting the typeface with the effect of RGB-split and an abrupt glitch that fosters the immersive feeling for the viewers.

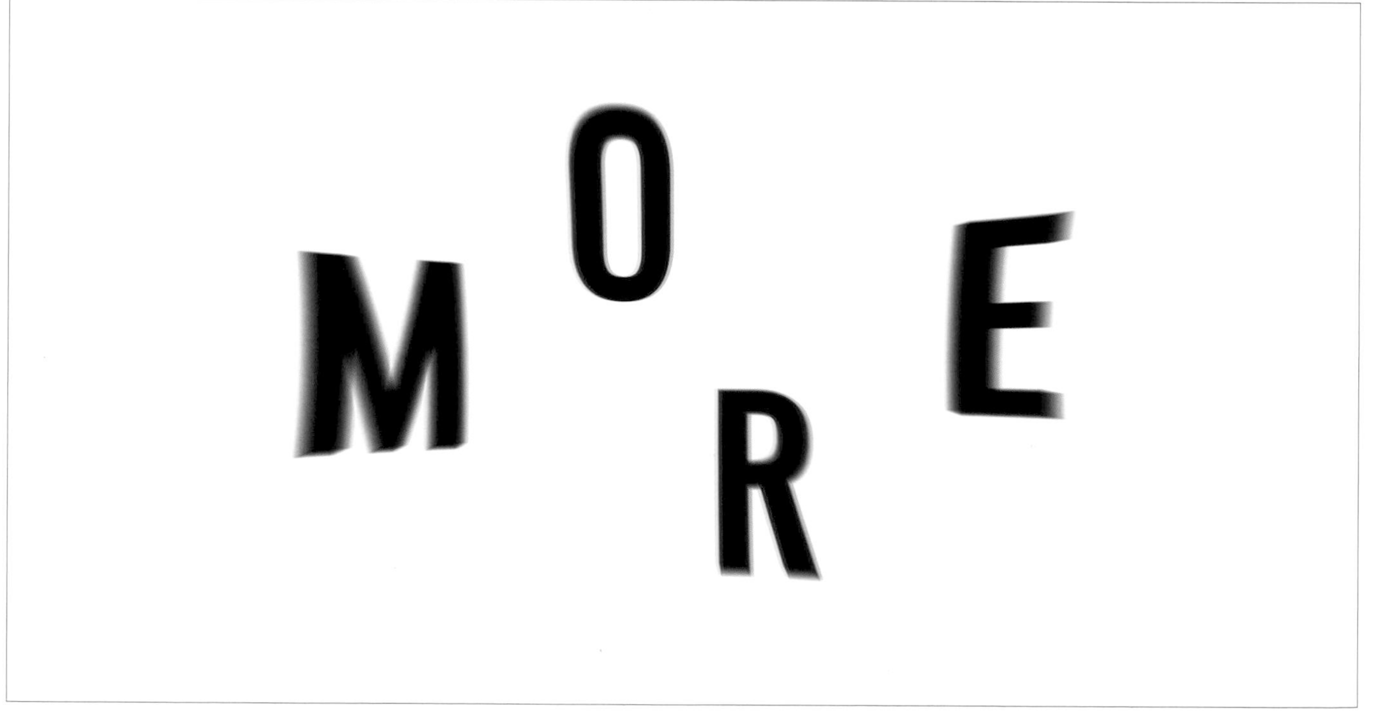

Vimeo

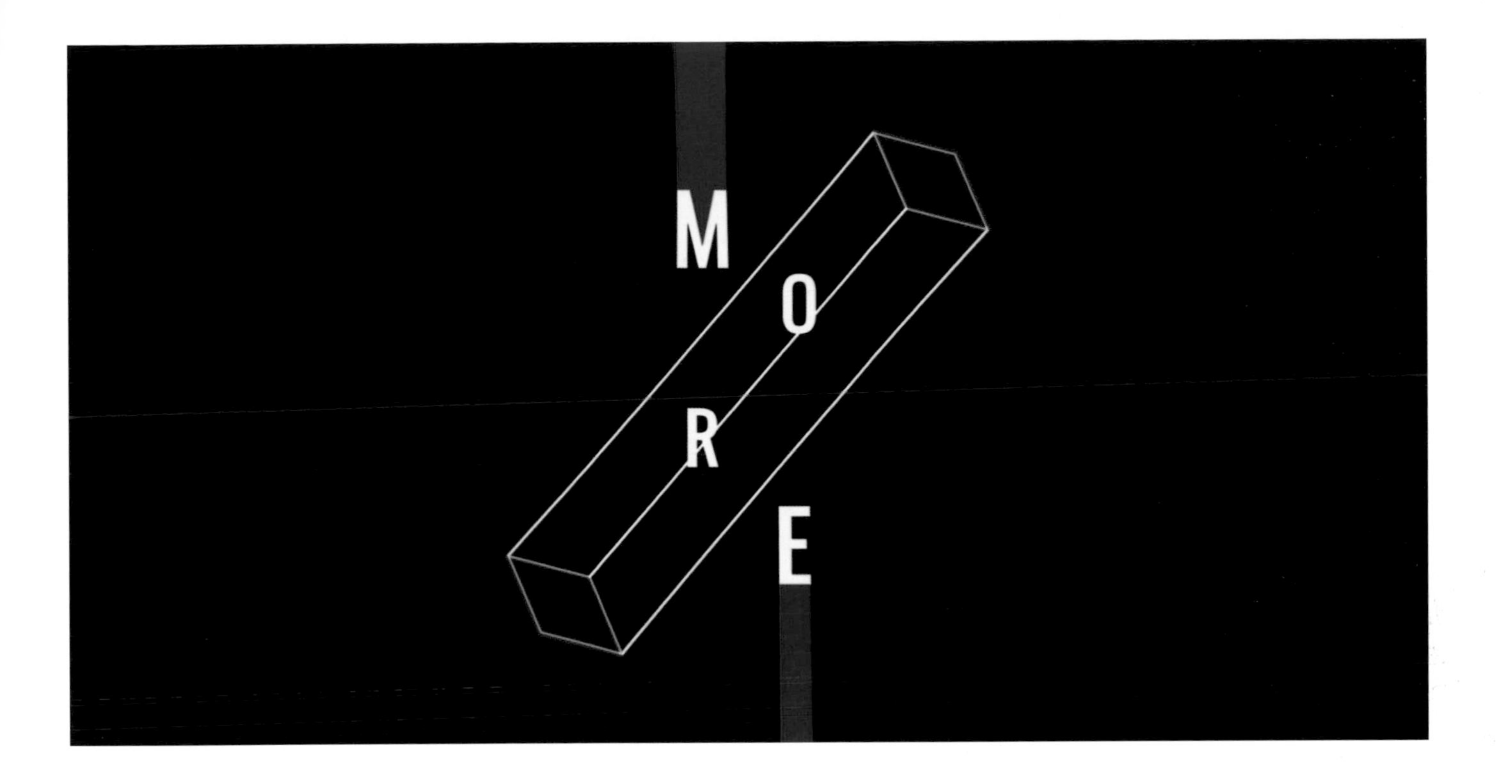

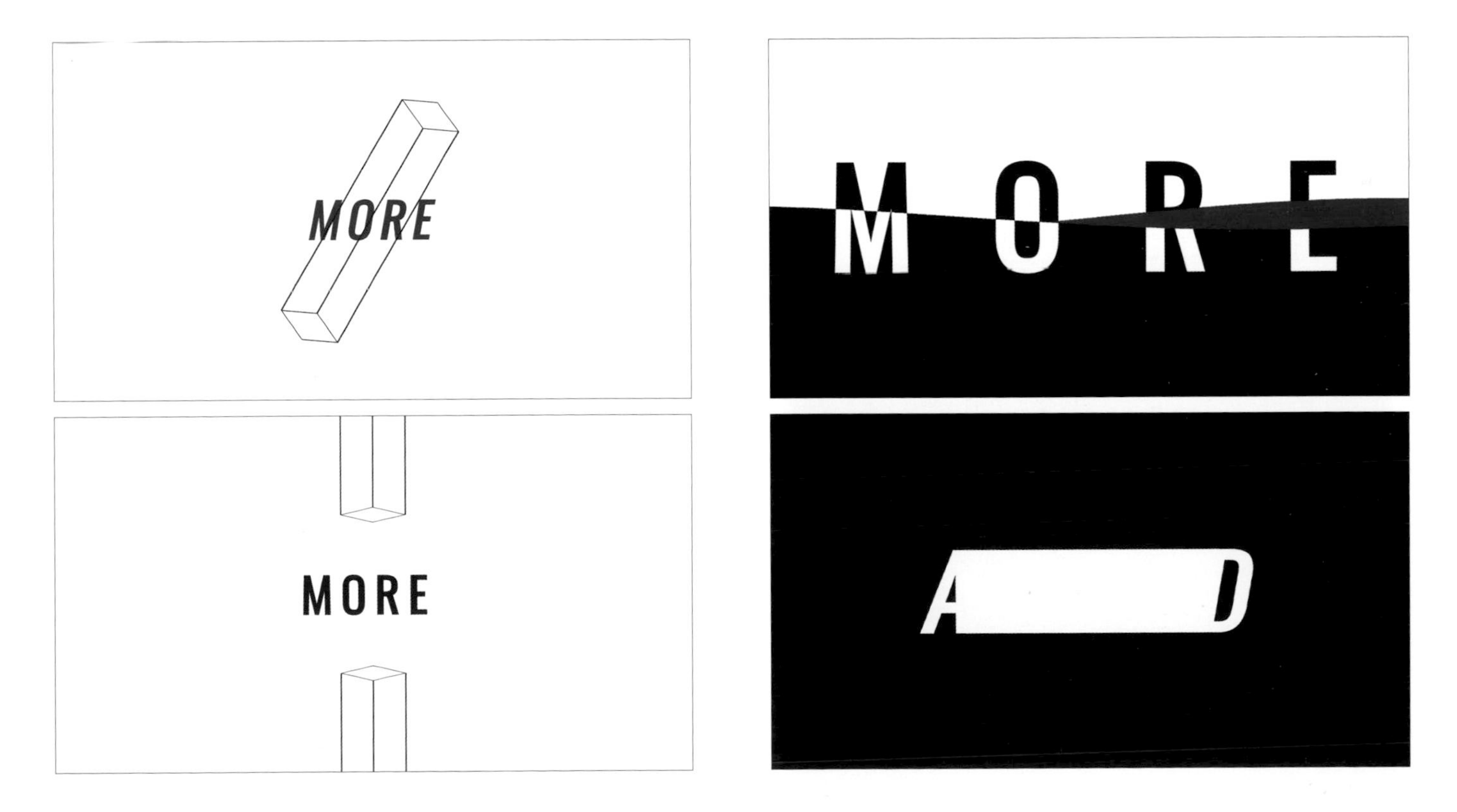

Independent Theatre Display Fonts

BY Moodagency

This project is the result of Moodagency's cooperation with The Progress Scene—an independent theatre located in Poznań, Poland. The main creative concept is to create one display font per every theatre season from December 2017 to January 2019 and use it on all formats of that season exclusively. Newly created fonts are able to transform and morph freely. So that slogans with such creative fonts can be responsive to various platforms—from prints to multiple devices and screen resolutions.

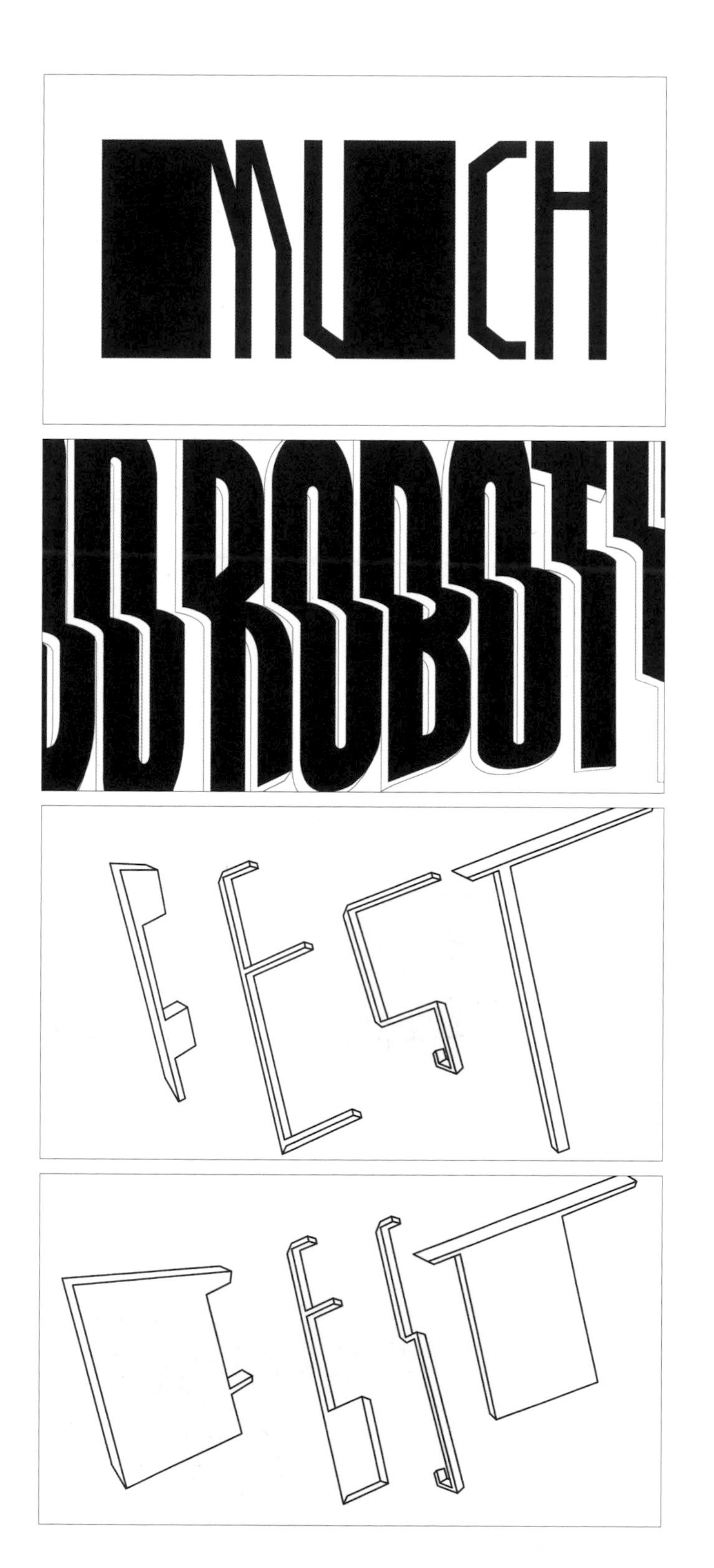

ART DIRECTION & DESIGN **Jakub Haremza** / MOTION DESIGN **Zosia Mikołajczak**

Behance
Official

display:
wladca
brochures
book
social ads
motion design
WSZ4SC4
RAZEM
DOM
MUCH

IBM 5 in 5: Making the Invisible Visible

BY Justin Au

Every year IBM publishes a report called "5 in 5"—five predictions of technological innovations in the next five years. The theme in 2017 was "Making the Invisible Visible." Each prediction focuses on a current "invisibility" and lays out how IBM will help make it visible. Inspired by optical illusions and optimism for the future, the six-piece animated graphics series interprets themes touching on environmental impact, healthcare, and language. As each piece starts as a sea of unknown, the animated typography and illustration uncover it to reveal each breakthrough.

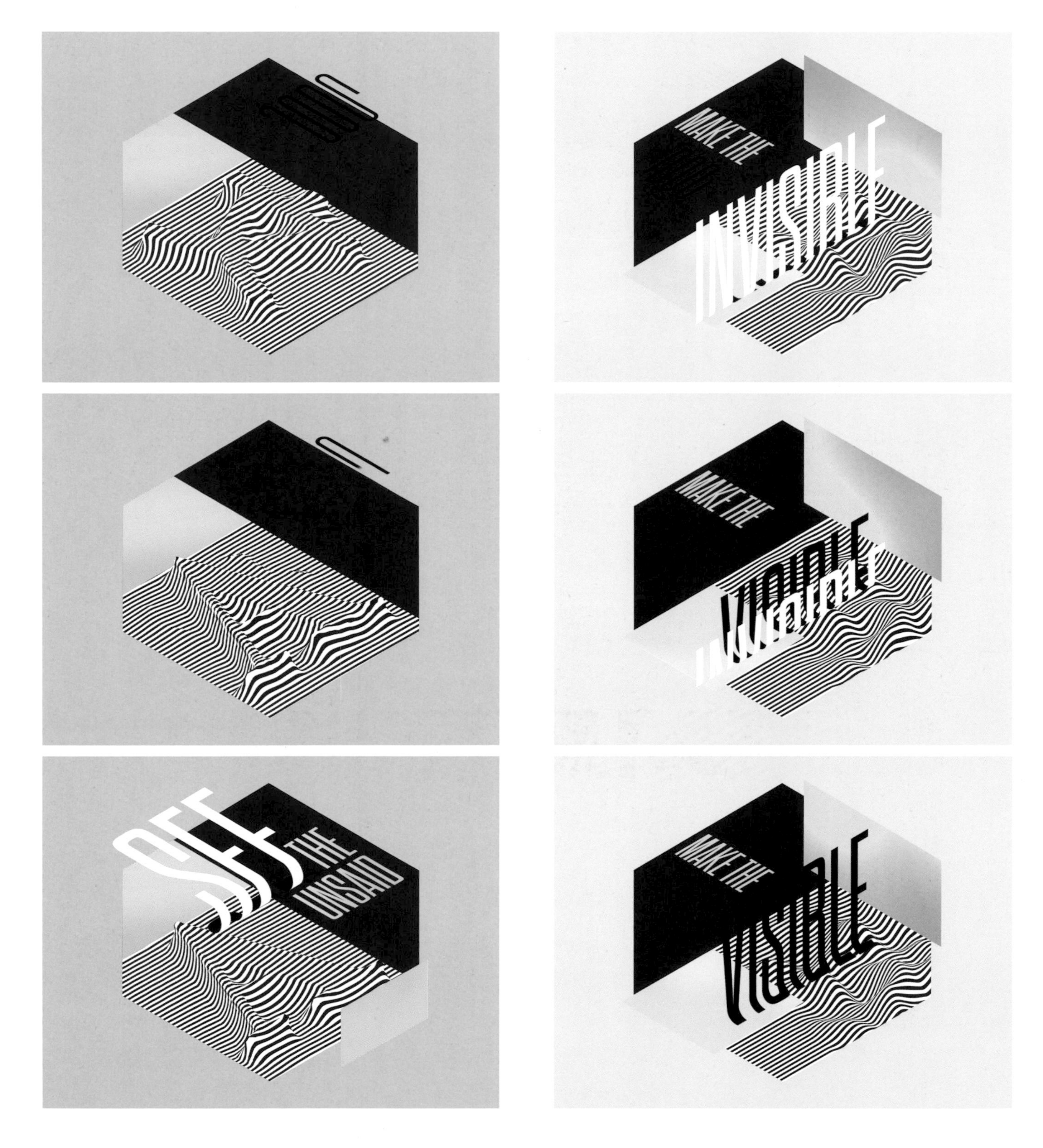

ART DIRECTION **Alvaro Masa** / CLIENT **IBM**

Behance

36 Days of Type

BY Vaidehi Vartak

This project, as a part of 36 Days of Type, explores the use of kinetic typography to create a dynamic interplay between type and colours. The animations, reminiscent of the Bauhaus Movement, follow an abstract pattern with various interpretations possible with one final objective—a playful harmony between the design elements.

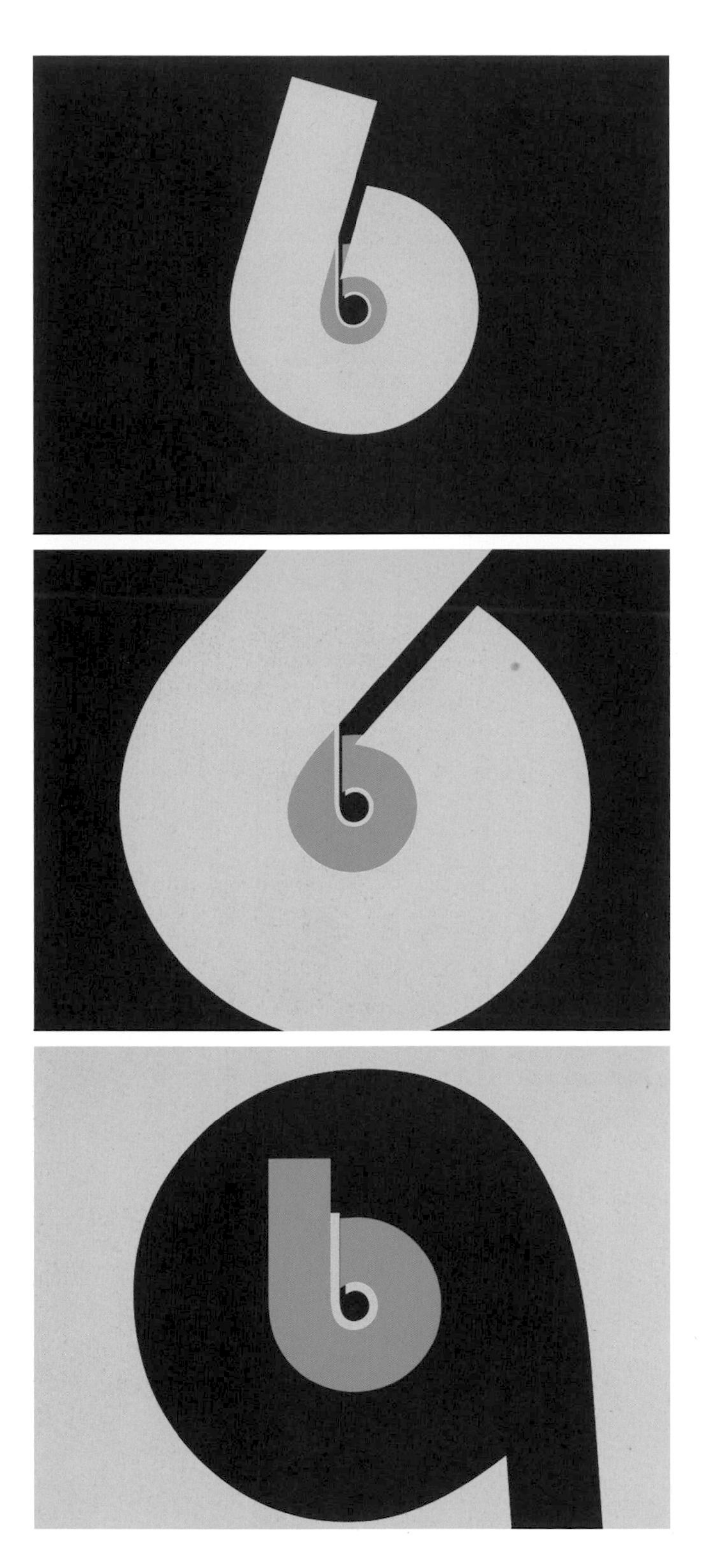

Dribbble

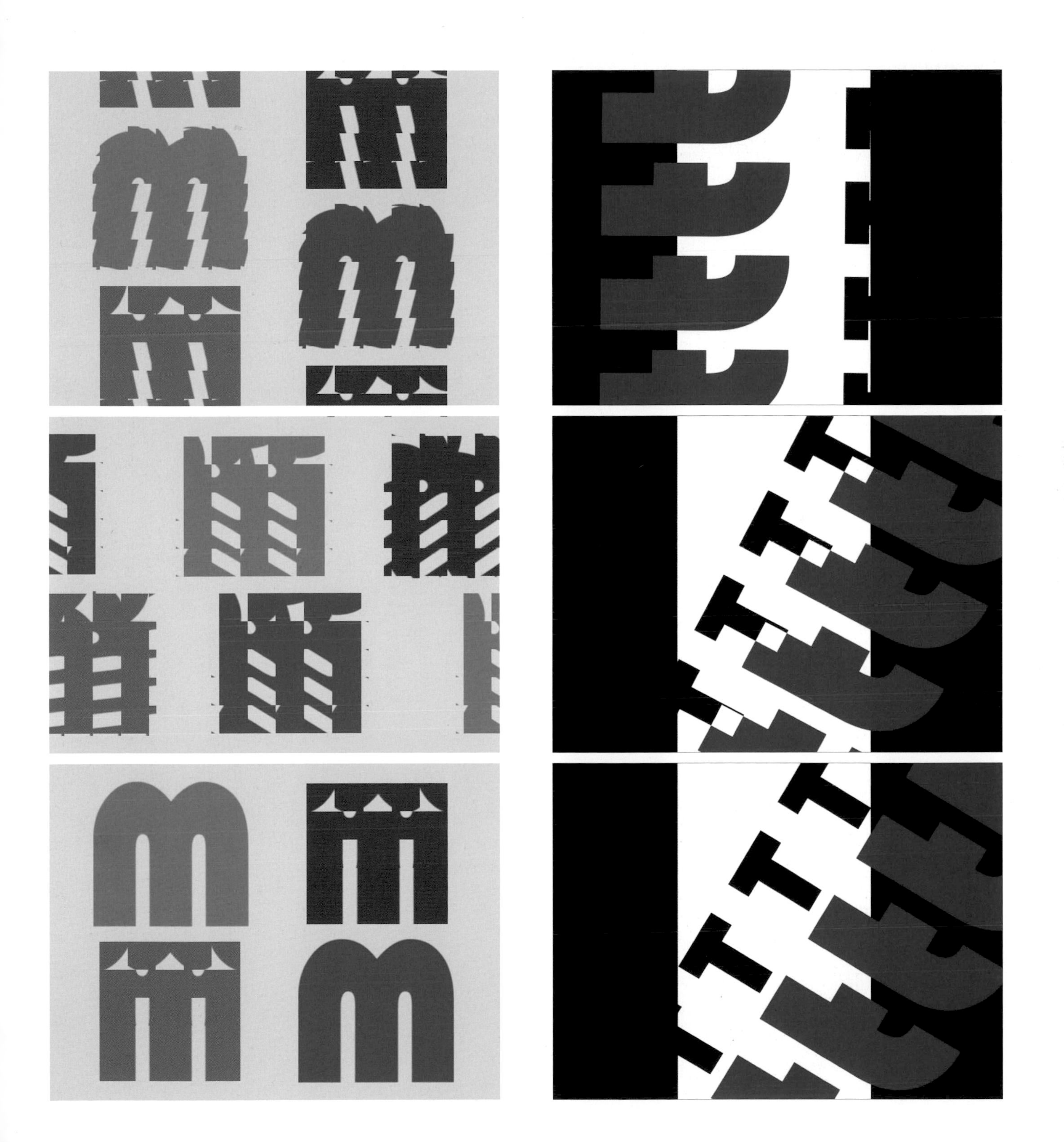

36 Days of Type

BY Andrius Tamosaitis

Andrius Tamosaitis started this project to repurpose his designed lower thirds. Each letter's idea related to that day's letter for 36 Days of Type, such as G for Grid, E for Exclude. Andrius designed the letters by mixing his designed and animated elements. Andrius created all the letters with simple shape layers in Adobe After Effects.

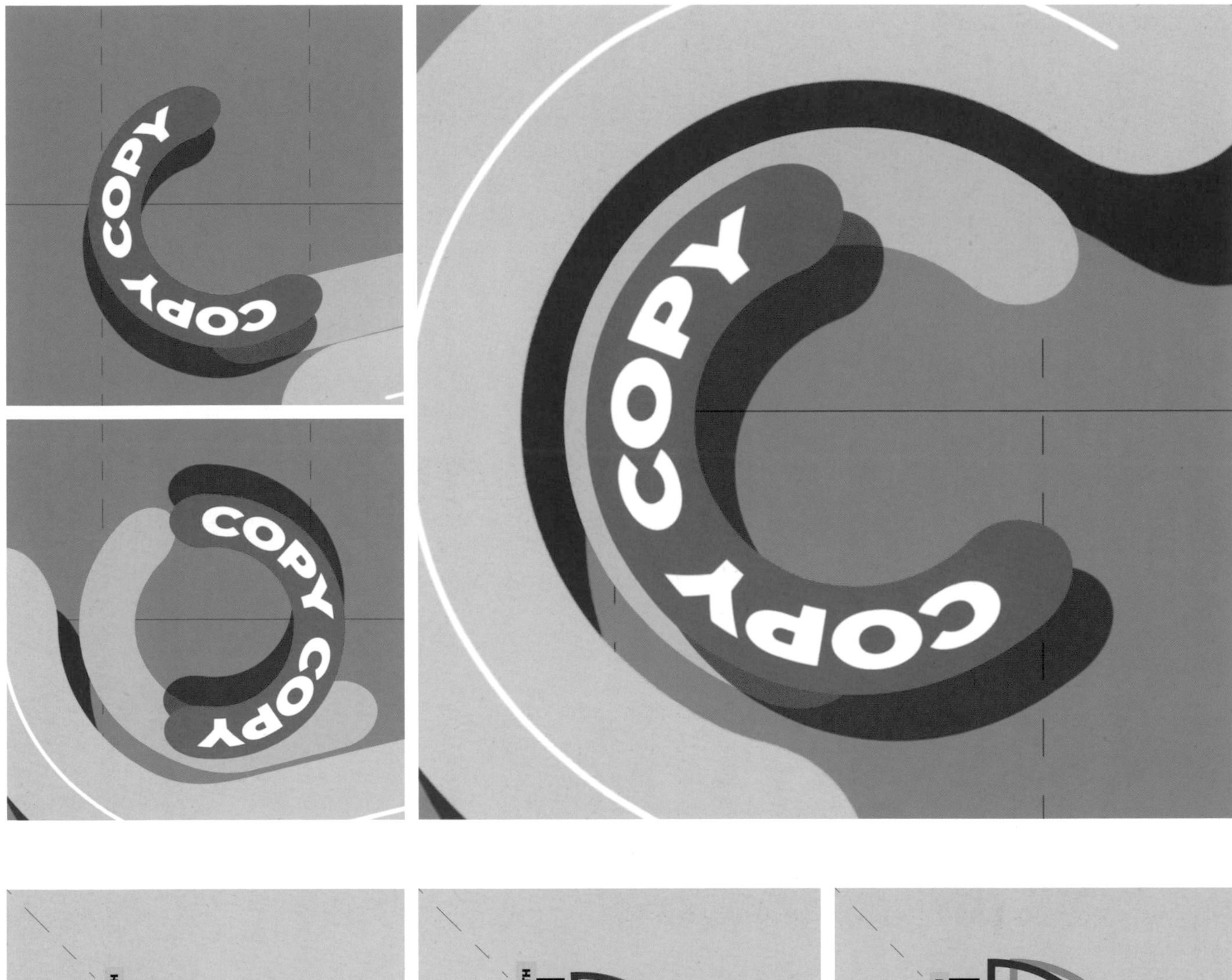

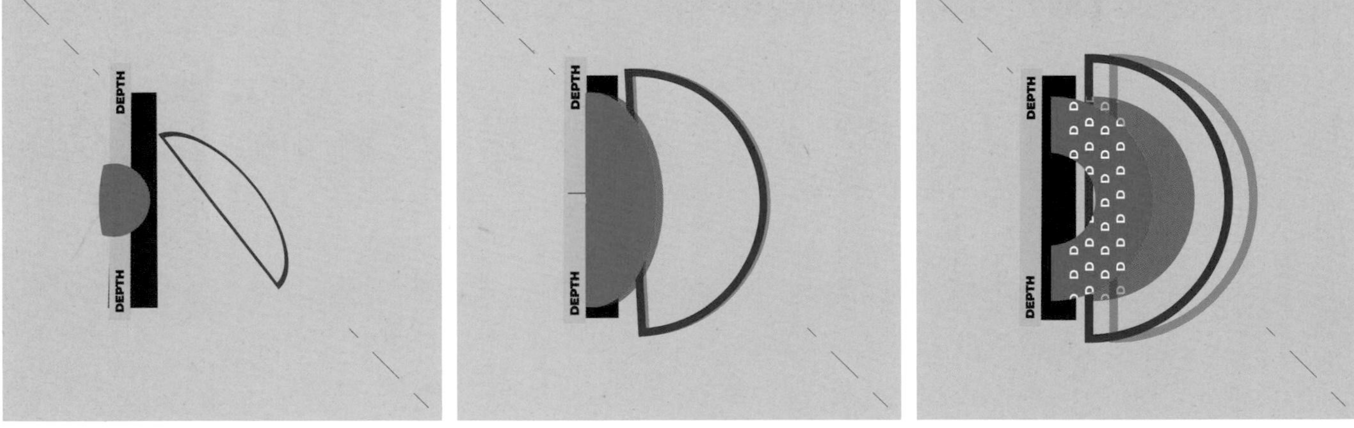

ANIMATION **Andrius Tamosaitis**

Dribbble

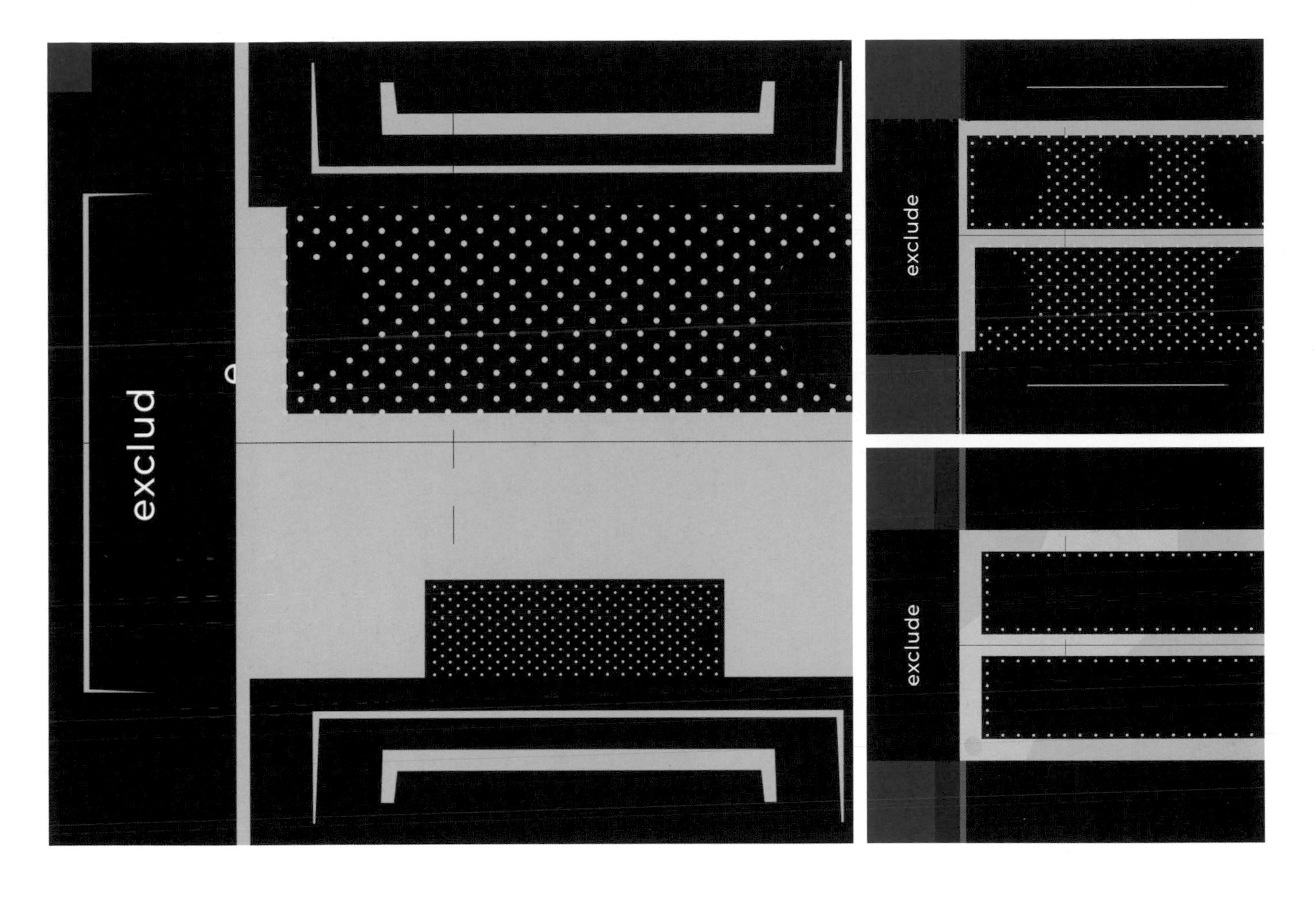

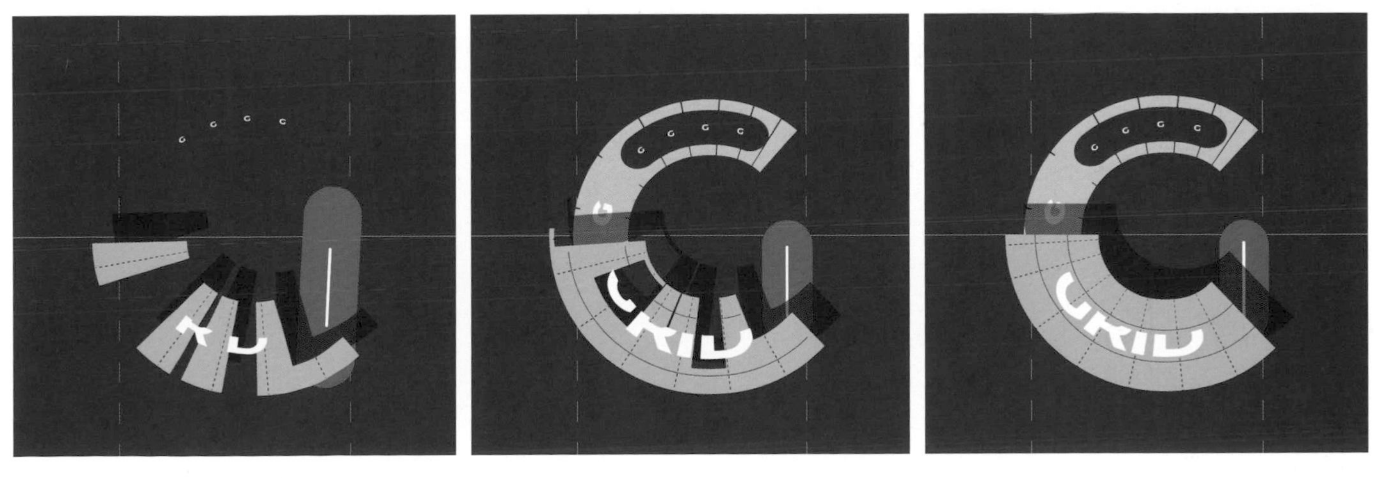

Typografik

BY Styleforces

Styleforces made this motion in Adobe After Effects. Styleforces wanted to expand the boundary of the software and explore 3D typography. The main goal was to keep motion simple and clean. Besides, their main inspiration is Swiss typography posters.

DESIGN **Michał Cineczny, Janusz Bartosiak**

Youtube

Animated Chinese Characters

BY Chen Junxian

In this project, Chen Junxian selected some of the common words from Classical Chinese. Some of the words are modal particles. Chen animated those words by nice matching colours and made a unique pace and visual style.

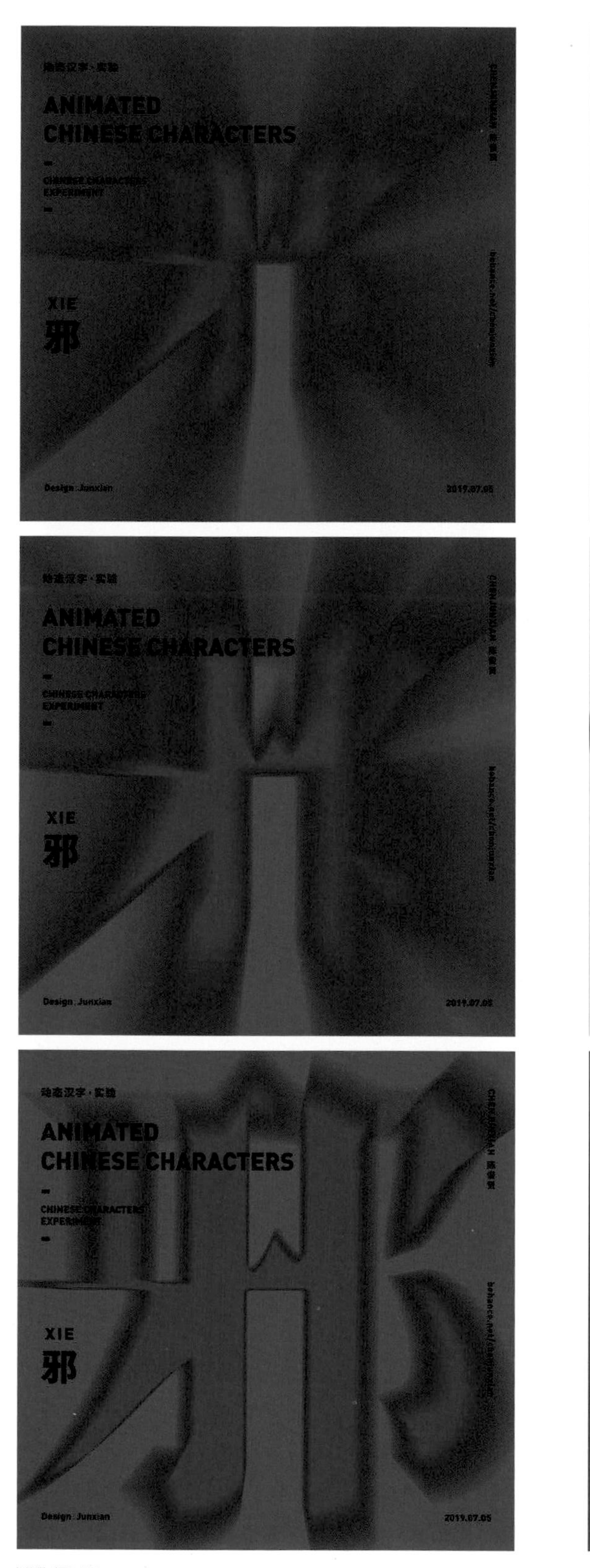

XIE (EVIL)

ZAI (ALAS)

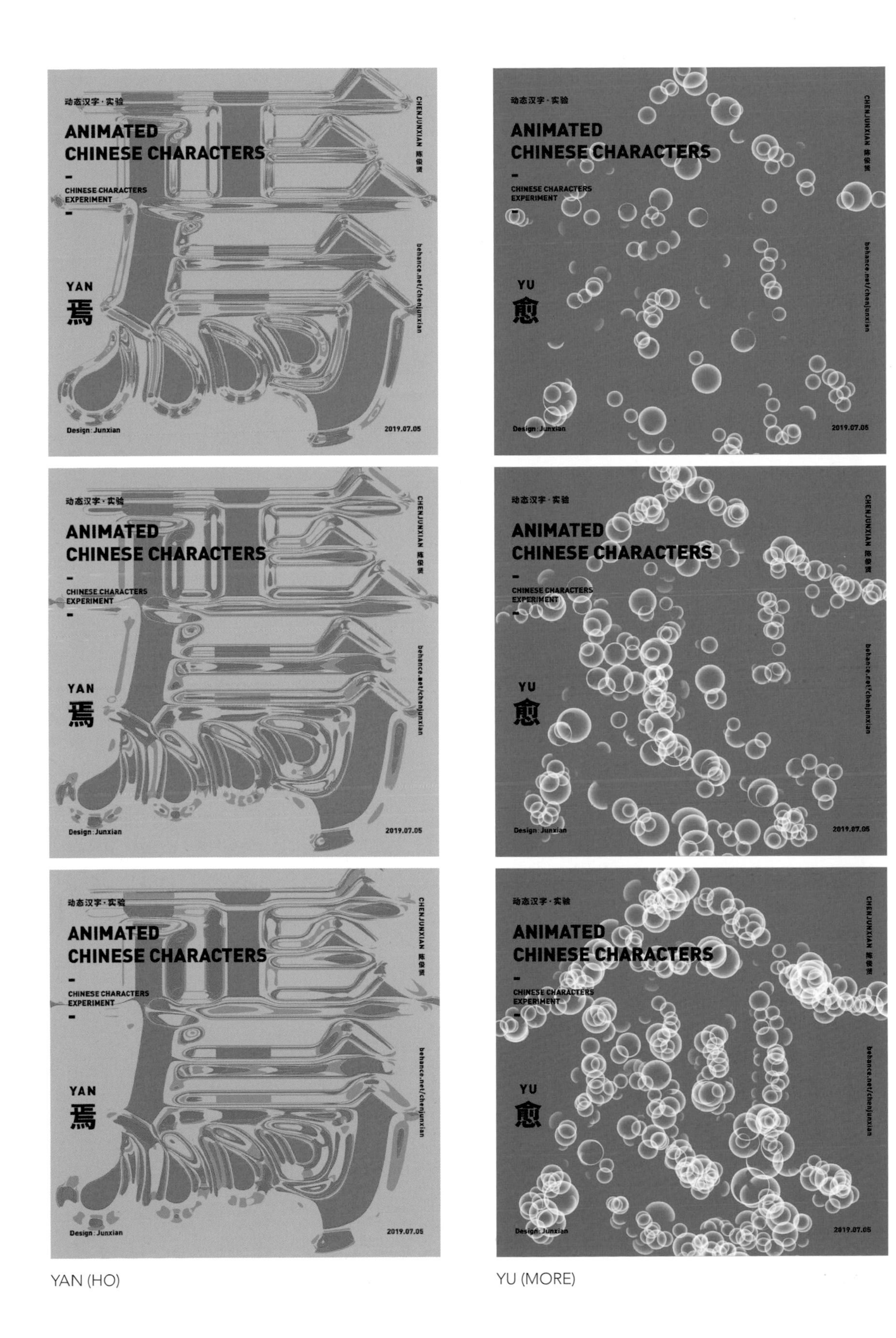

YAN (HO)

YU (MORE)

Fora de lloc. Sexual Assault Prevention Campaign

BY democràcia estudi

Faced with a sensitive issue such as the prevention of sexual aggression in nightlife, democràcia estudi wants the messages to be the key visual of the campaign, without images that divert the attention of the viewers. For this, the designers used striking colours and kinetic typography in the messages, which allow them to achieve the established objective.

DESIGN **Javi Tortosa, Migue Martí, Chavo Roldán** / COPYWRITING **Marta Tortosa, Pablo Llobell** / CLIENT **Ajuntament de València**

Behance Official

Type in Motion Vol. One

BY Holke79

This project consists of 2D and 3D animated typographic experiments. Holke79 stated that one of the most fascinating and enjoyable tasks for motion design are to animate the texts and typography. To pay tribute to the flashing signs with texts on the streets in his childhood, Holke79 created these loop animations with typography by Cinema 4D and Adobe After Effects. Holke79 hopes that some pieces in this project can be shown on the streets.

Behance

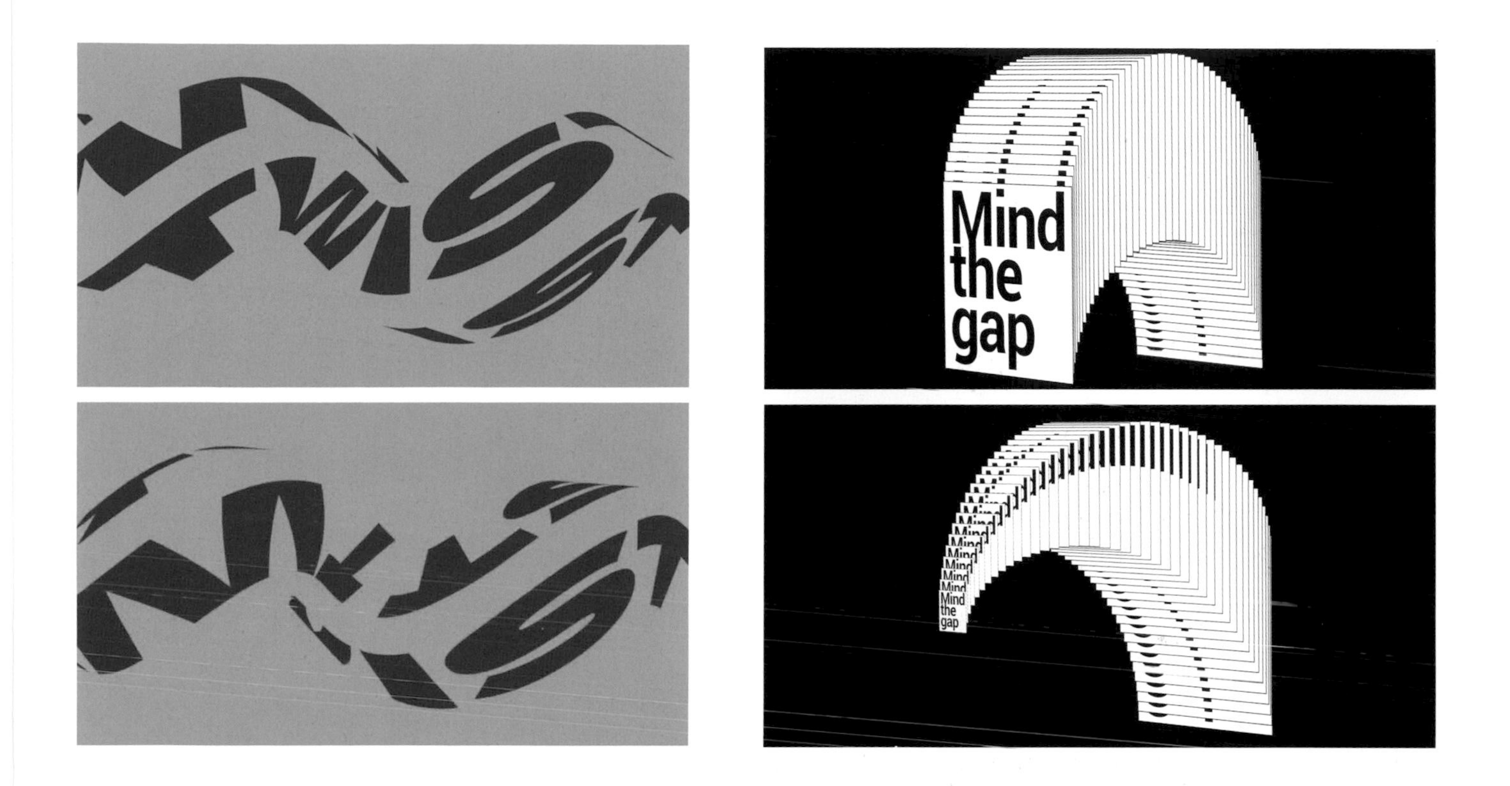
Mind
the
gap

finish just
it. finish

H
36DaysOfType

U
36DaysOfType

H
36DaysOfType

U
36DaysOfType

O
36DaysOfType

P
36DaysOfType
P
36DaysOfType

O
36DaysOfType

Kinetic Type Exploration

BY Roberto Warner

This project explored the typography on screen and the potential replacement of printed posters. As screens have become more prevalent, typography can push the boundary beyond the limitations of print and express its dynamics on screens. Roberto Warner used a bold typeface to express different words and ideas. And exploring the relationship between movement and the text's connotation is the motivation behind this project.

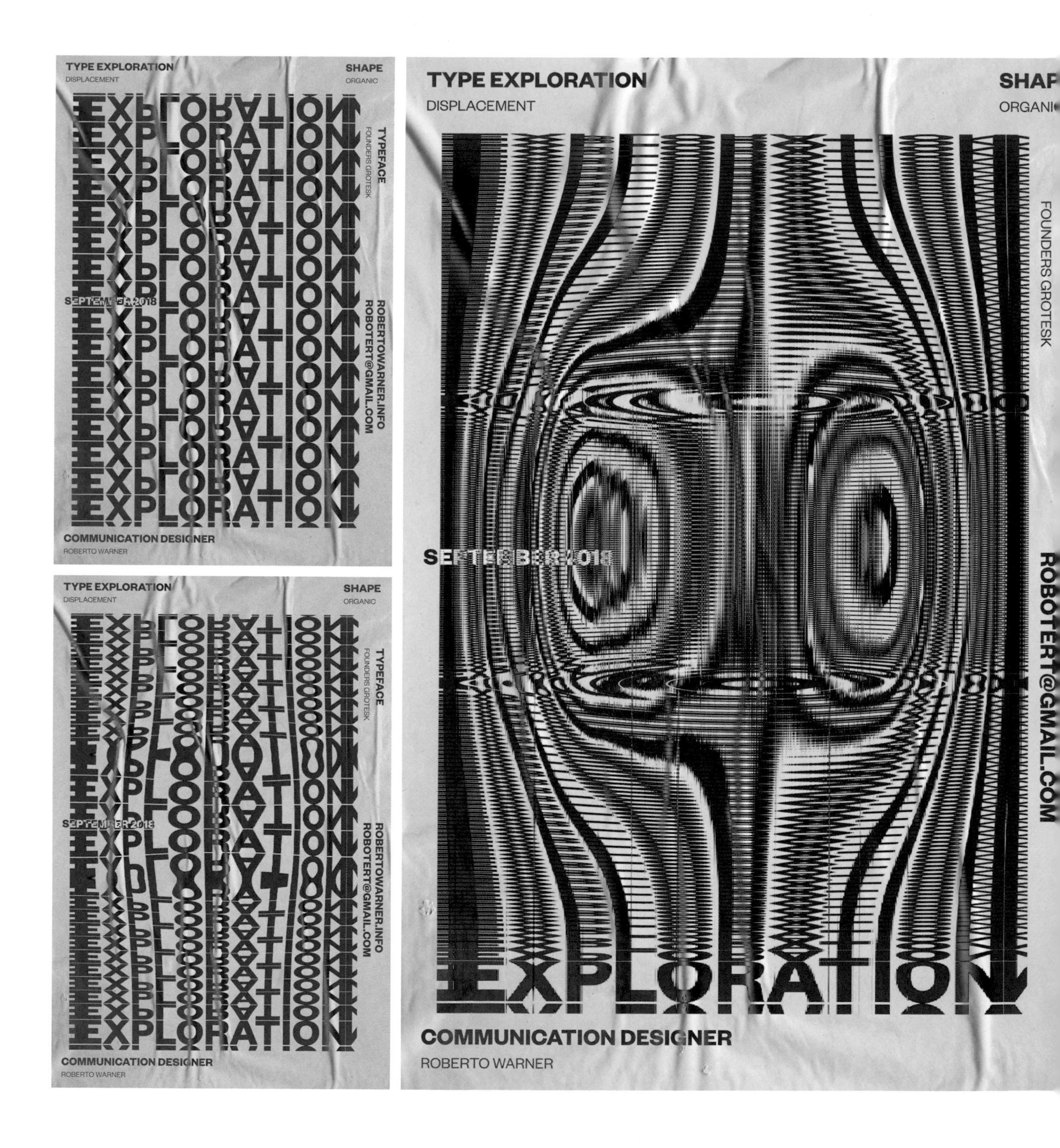

CREATIVE DIRECTION **Roberto Warner**

Behance

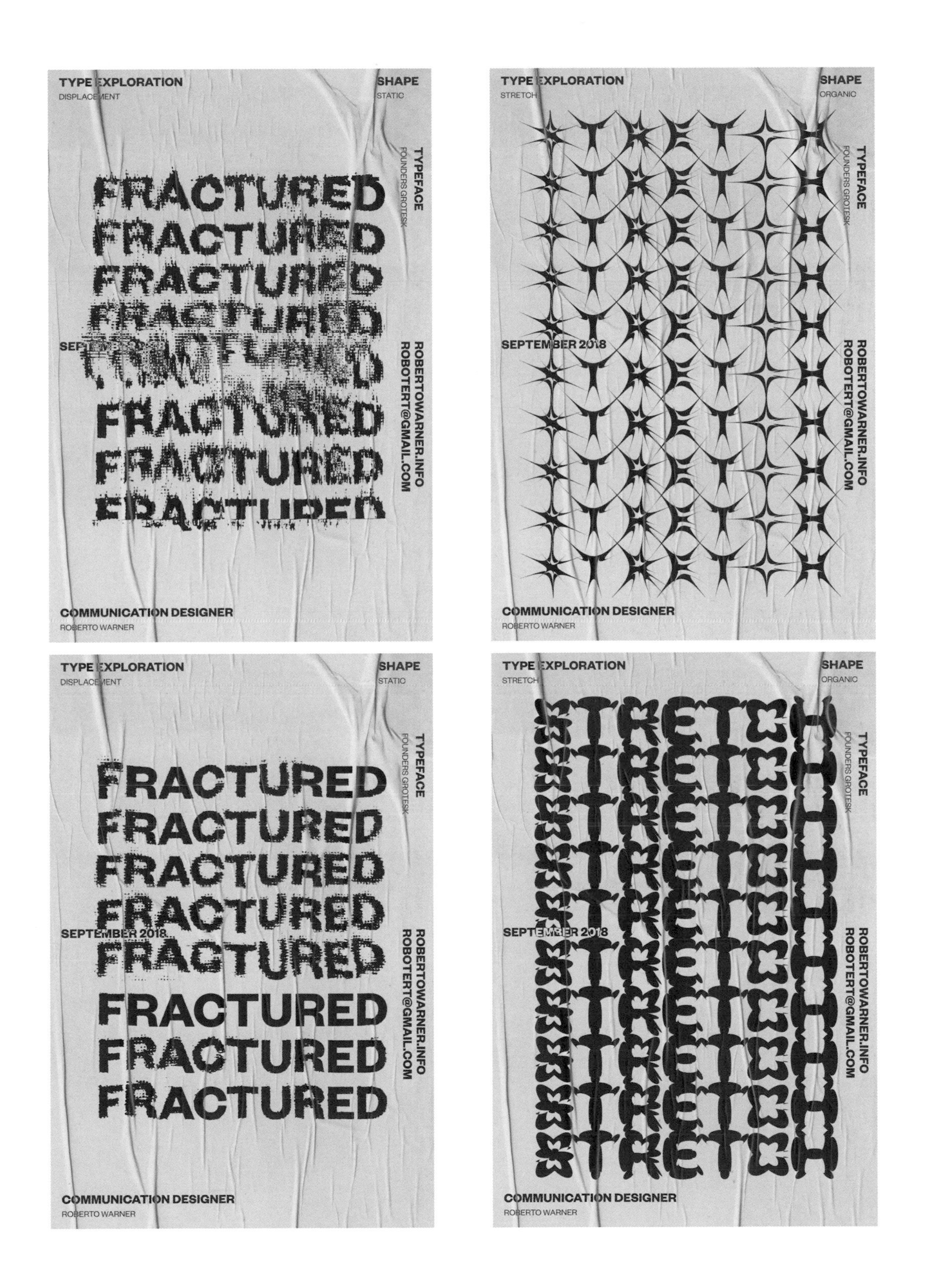
TYPE EXPLORATION
DISPLACEMENT
SHAPE
STATIC
TYPEFACE
FOUNDERS GROTESK
FRACTURED
FRACTURED
FRACTURED
FRACTURED
FRACTURED
FRACTURED
FRACTURED
FRACTURED
SEPTEMBER 2018
ROBERTOWARNER.INFO
ROBOTERT@GMAIL.COM
COMMUNICATION DESIGNER
ROBERTO WARNER
TYPE EXPLORATION
STRETCH
SHAPE
ORGANIC
TYPEFACE
FOUNDERS GROTESK
SEPTEMBER 2018
ROBERTOWARNER.INFO
ROBOTERT@GMAIL.COM
COMMUNICATION DESIGNER
ROBERTO WARNER

Animated Chinese Characters

BY Hunk Xing

Hunk Xing tries to express the meaning of the Chinese characters by animations. For example, the character Ai (Cancer) in animated form represents the cancer cells' invasion in the human body with a linear process from healthy to death. With the sound effect referring to morbid change, the mood of the animation becomes viewable.

Basing on the design concept and process of the animated character Ai (Cancer), the whole series follows a design principle—a combination of motion, quietness, and spirit that multidimensionally communicates the meanings of Chinese characters and their own charm.

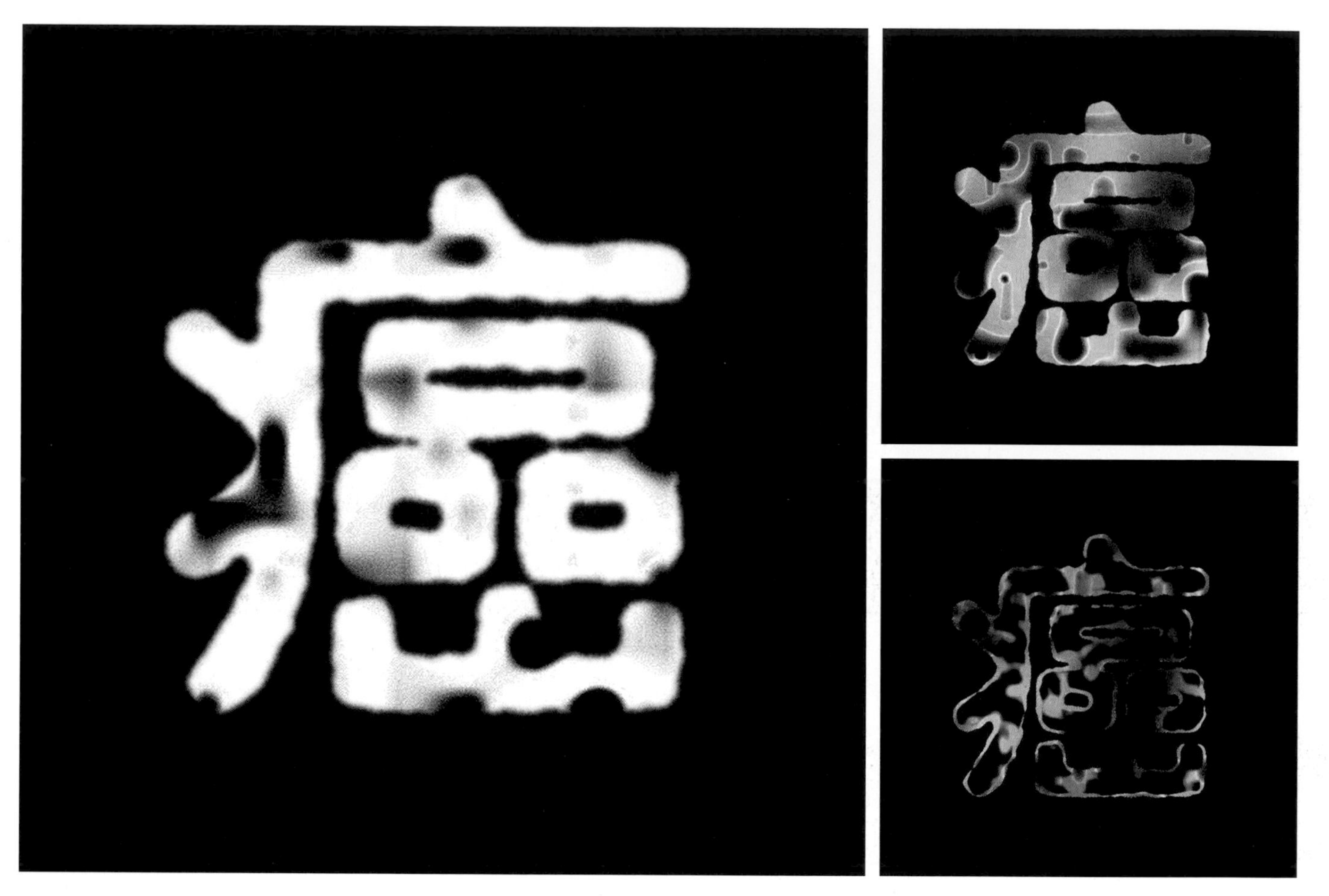

▴ AI (CANCER)

▾ YING (SHADOW)

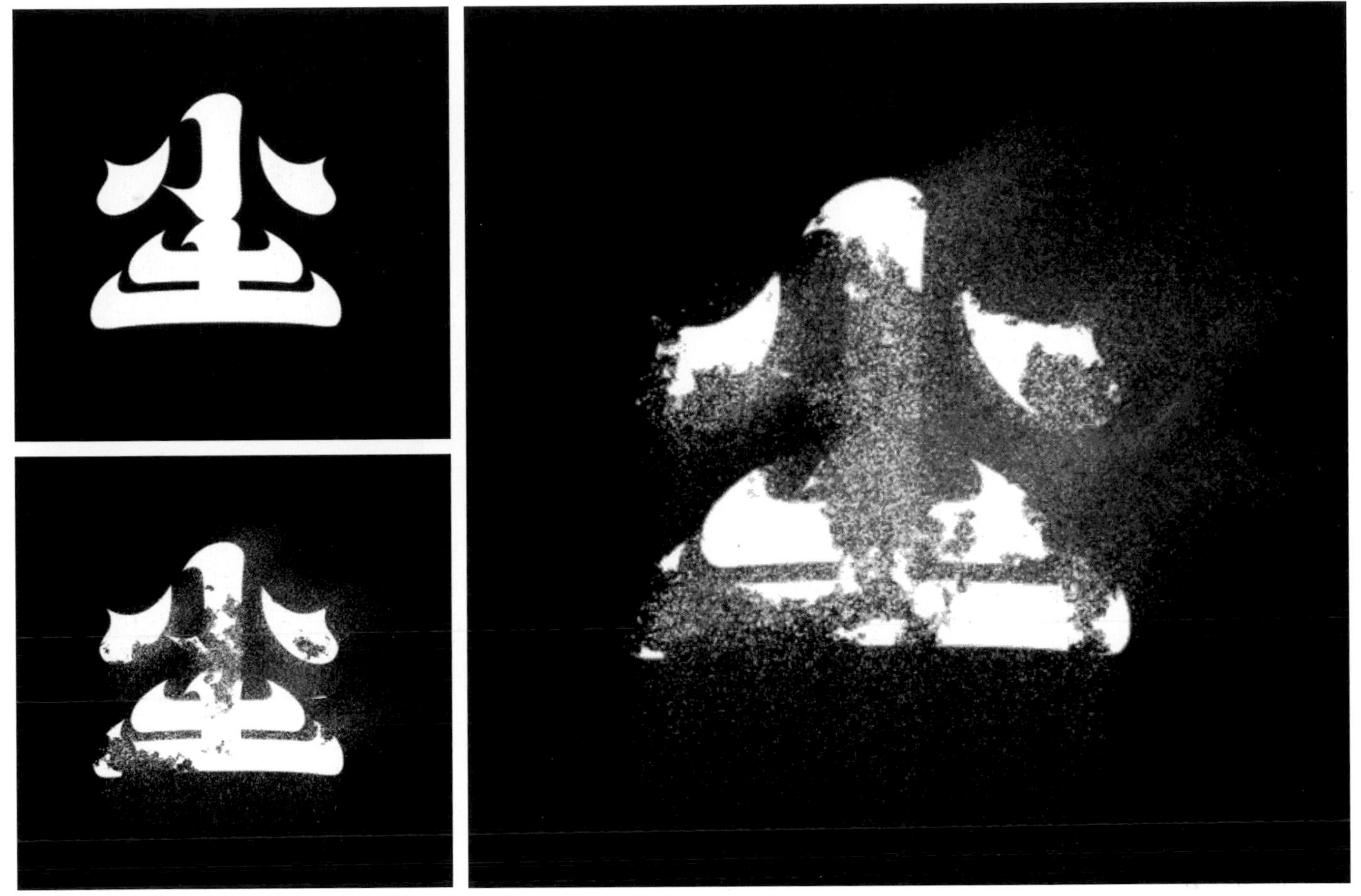

▲ CHEN (DUST)
▼ PIAO (FLOAT)

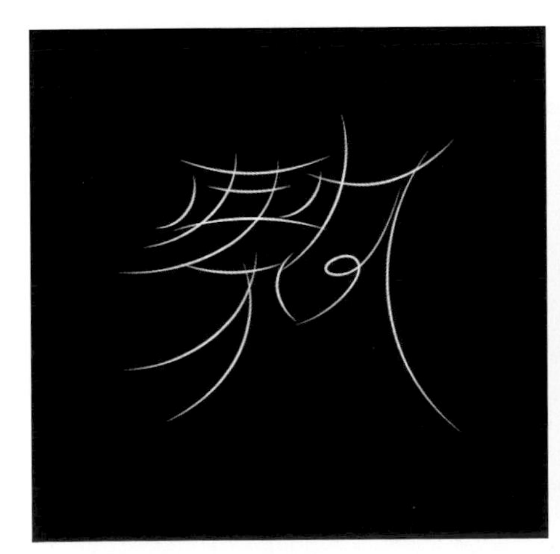

36 Days of Type

BY Nuno Leites

This is a collaborative project between Estaminé Studio and Nuno Leites which consists of creating a letter and figure based on the three basic shapes: square, circle, and triangle. Estaminé Studio did the substantial papercut for each letter and figure. And Nuno simultaneously transformed the papercut into motion graphics.

ART DIRECTION & CREATIVE DIRECTION **Estaminé Studio, Nuno Leites** / MOTION DESIGN **Nuno Leites**

Behance

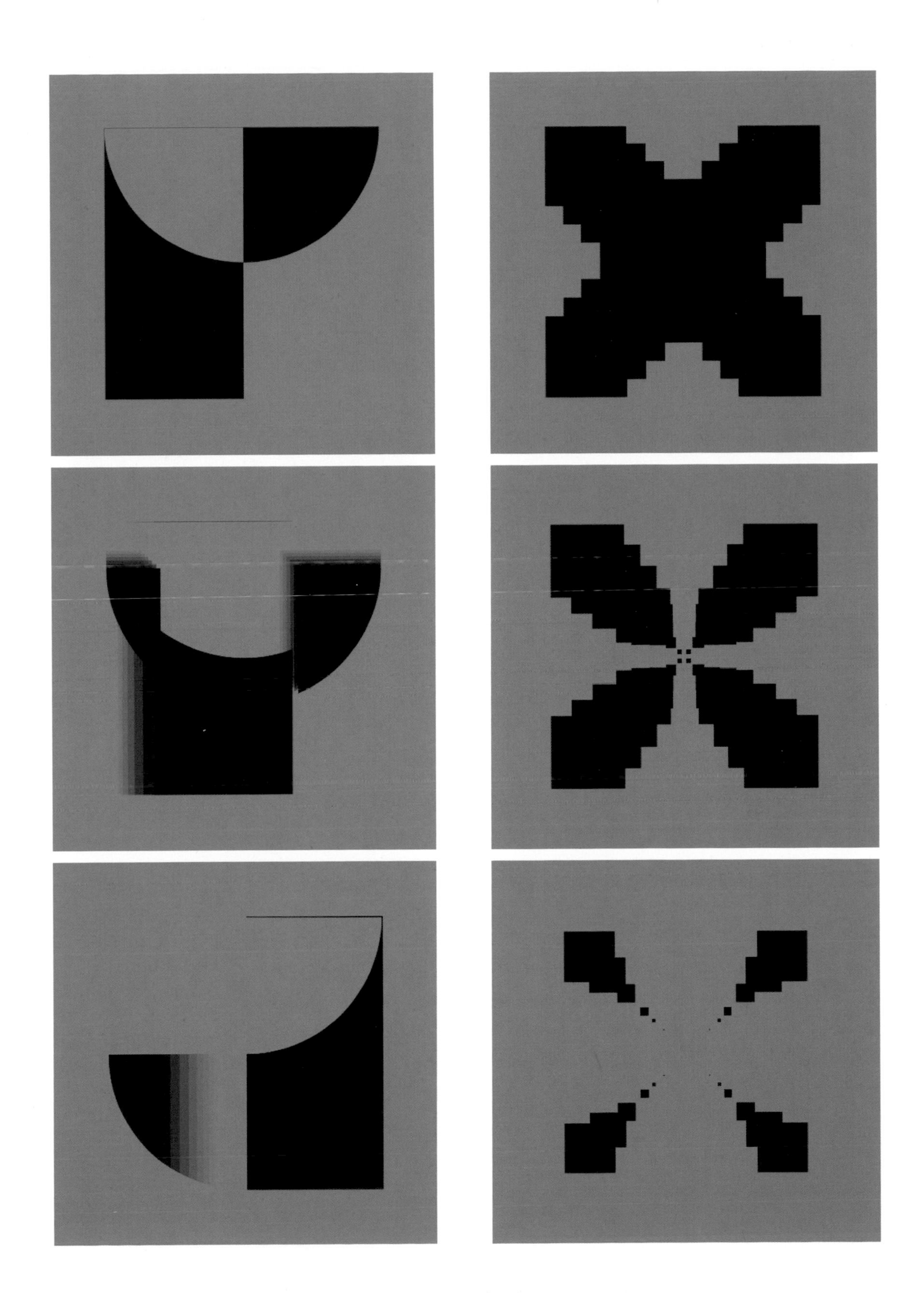

Fire Font

BY Kid Pixel

This handmade font was designed by Kid Pixel (Camille Moine), which can heat the viewers' imagination to the flashpoint. The letters of fire font resemble the candles slowly burning in the dark. What is more, The appearance and animation of the Fire Font reflect some kinds of paradoxical feeling, both encouraging and thrilling, warm and mysterious, safe and dangerous.

Behance

MOTION DESIGN & TYPOGRAPHY DESIGN **Kid Pixel**

Motion Type Project

BY Ting-An Ho

This project concludes the motion design of Chinese typography. Because of the difference between oriental and western characters, Ting-An Ho realises that it was not entirely appropriate to copy the same design from western characters. Understanding the strokes and meaning of different characters, Ting-An Ho finally designed several motion designs on screen for evaluation of motion graphics for Chinese.

Behance

Tidy Font

BY Kid Pixel

Tidy Font represents geometric classic elegance. The animated letters move swiftly and elegantly, creating an inerasable look for the viewers. And the Tidy Font is practically a universal remedy for designs requiring precise lines and mathematical subtlety. With its futuristic appeal, Tide Font is also a good choice for sci-fi projects.

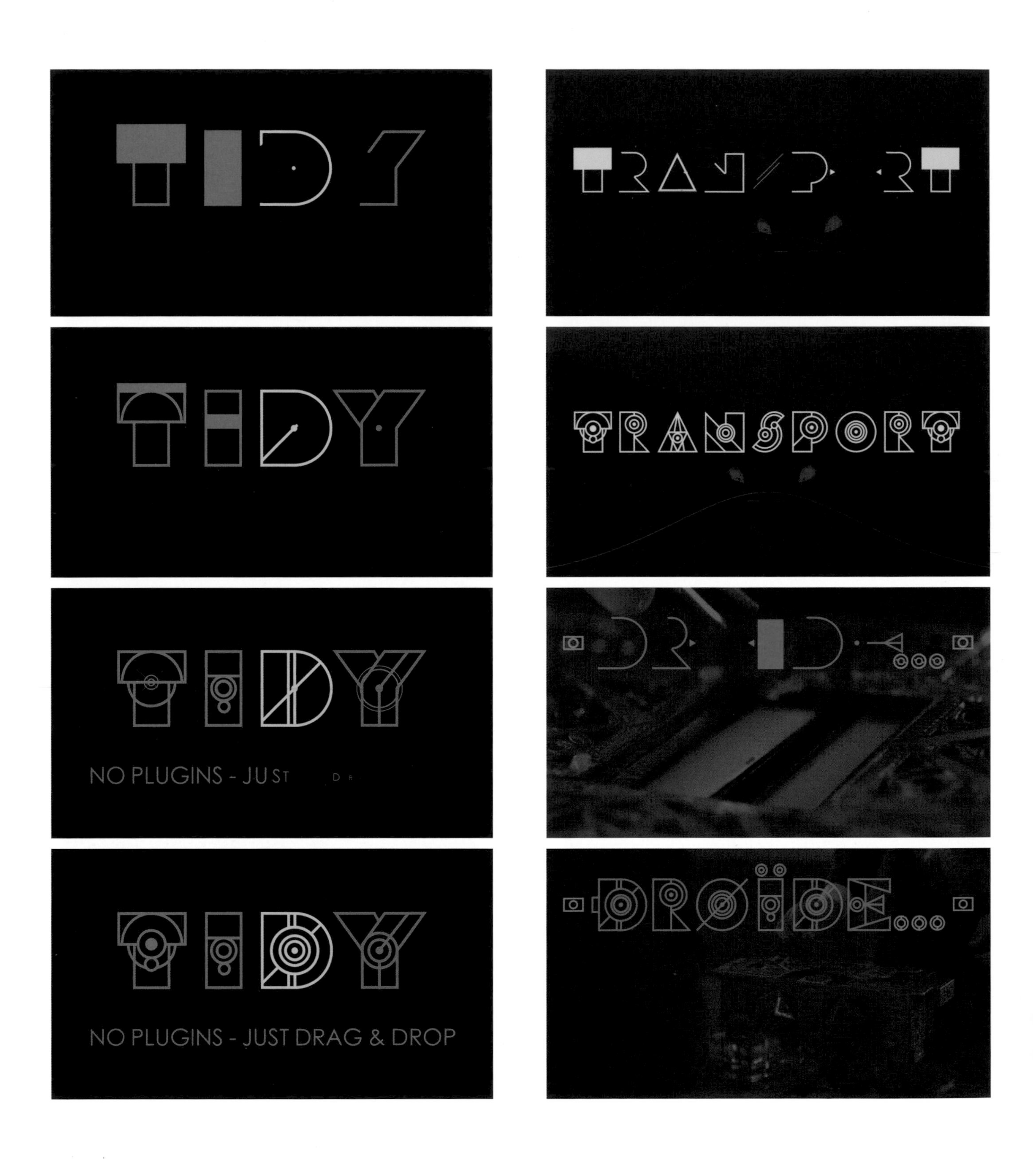

MOTION DESIGN **Kid Pixel** / TYPOGRAPHY DESIGN **Ryan Corey**

Behance

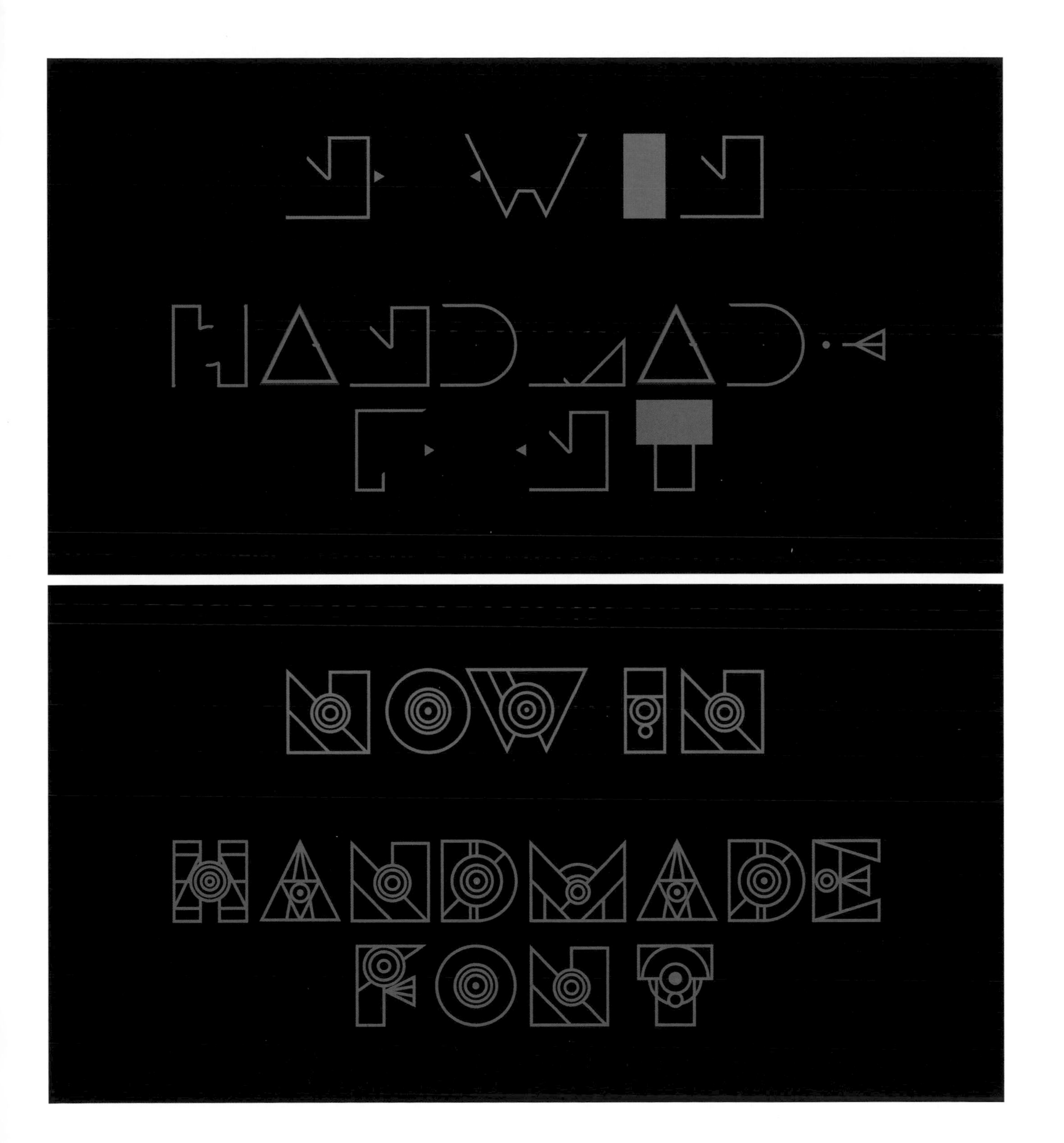

Fixture: A Kinetic Type Specimen

BY Alejandro Paul

Fixture is a massive font family taking on plentiful offerings of the late 19th century's typeface, posters, and wooden letterpress done in the Grotesk genre. This font family offers wide functional flexibility to work on different media, including film's credits. With the help of Vanessa Zuñiga, the kinetic and playful visual experiments enrich the imagination of the viewers.

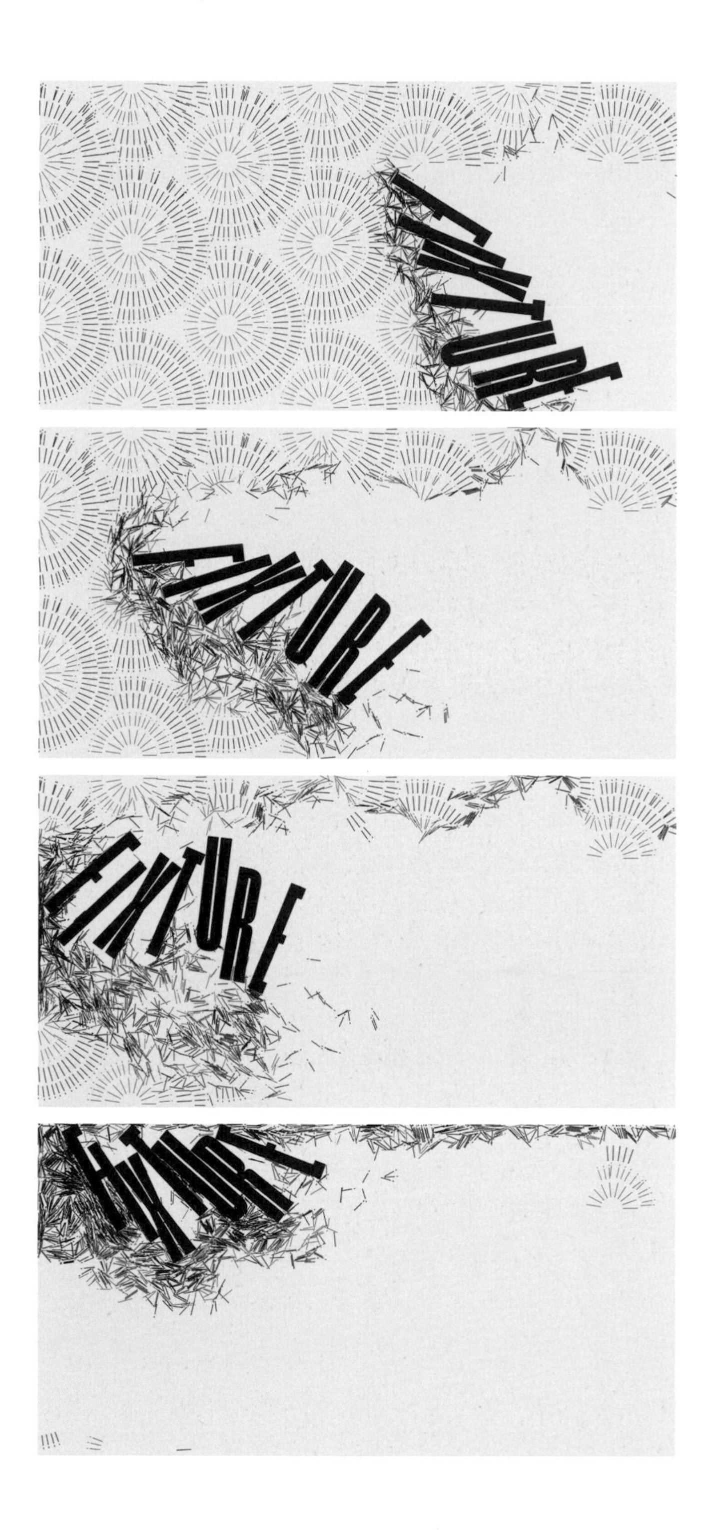

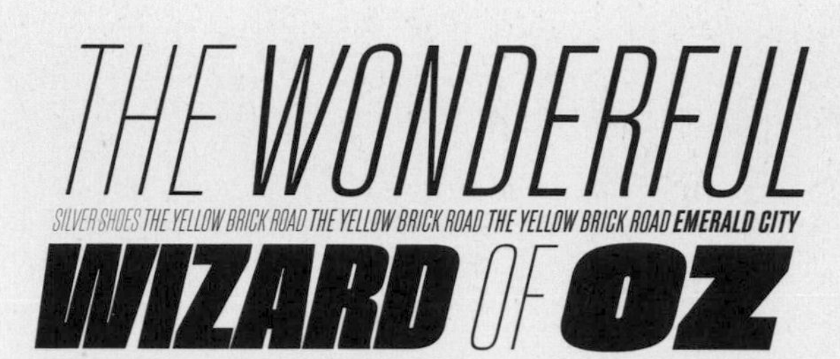

ART DIRECTION **Ale Paul** / ANIMATION **Vanessa Zuñiga**

Behance

WES
DERSON
QUATIC
with Steve Zi

Fixture
A new font by Sudtipos

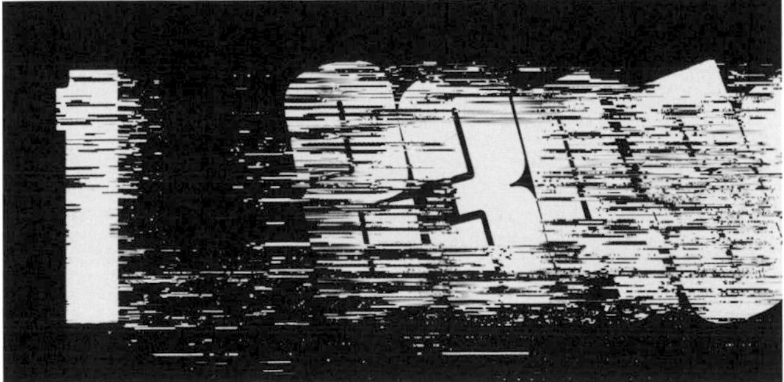

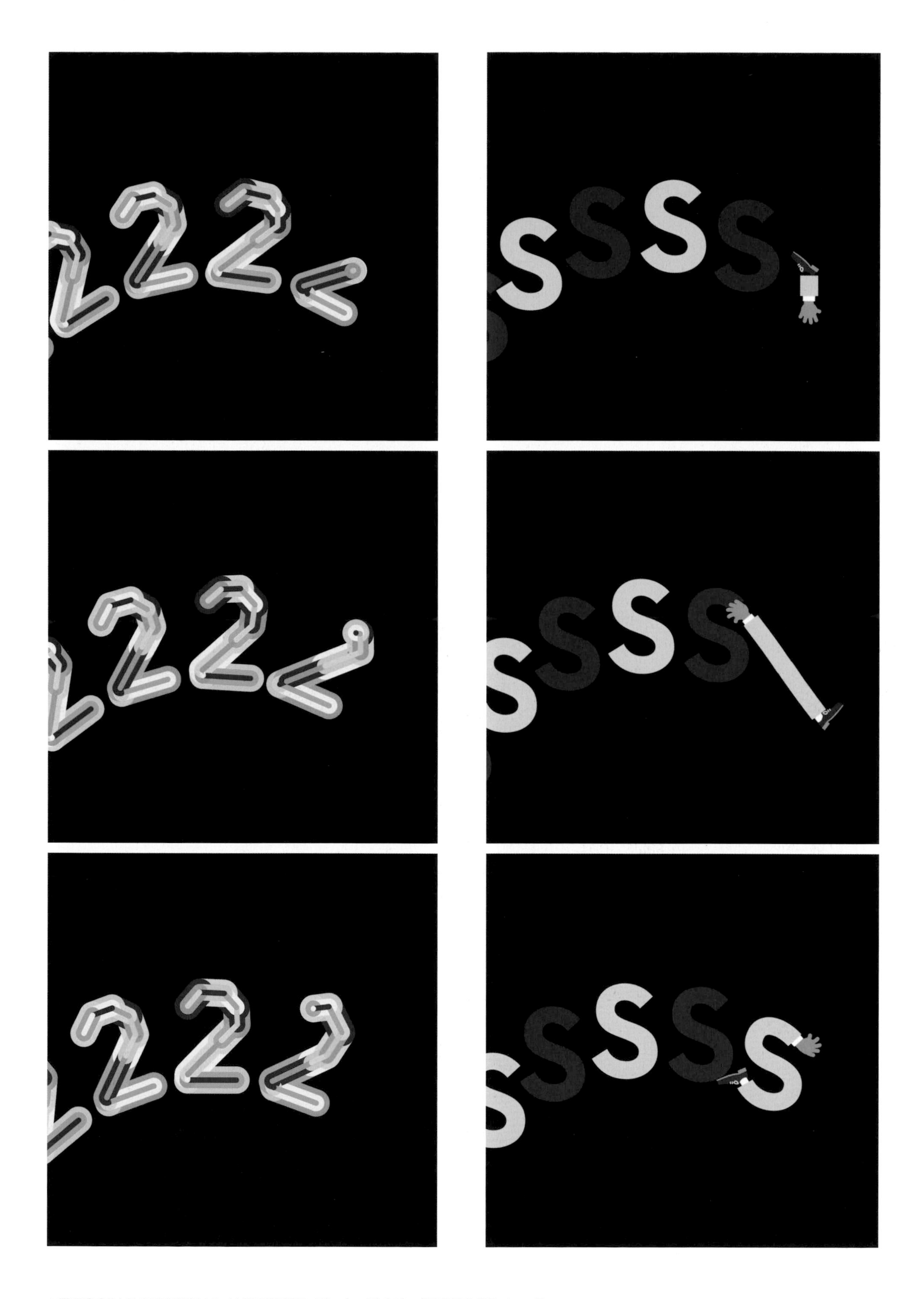

‹ TYPOGRAPHY DESIGN & ANIMATION **Nicolas Lichtle** / TYPEFACE **Anodine**

› TYPOGRAPHY DESIGN **Hanken Co.** / ANIMATION **James Curran** / TYPEFACE **Mobilo Bold**

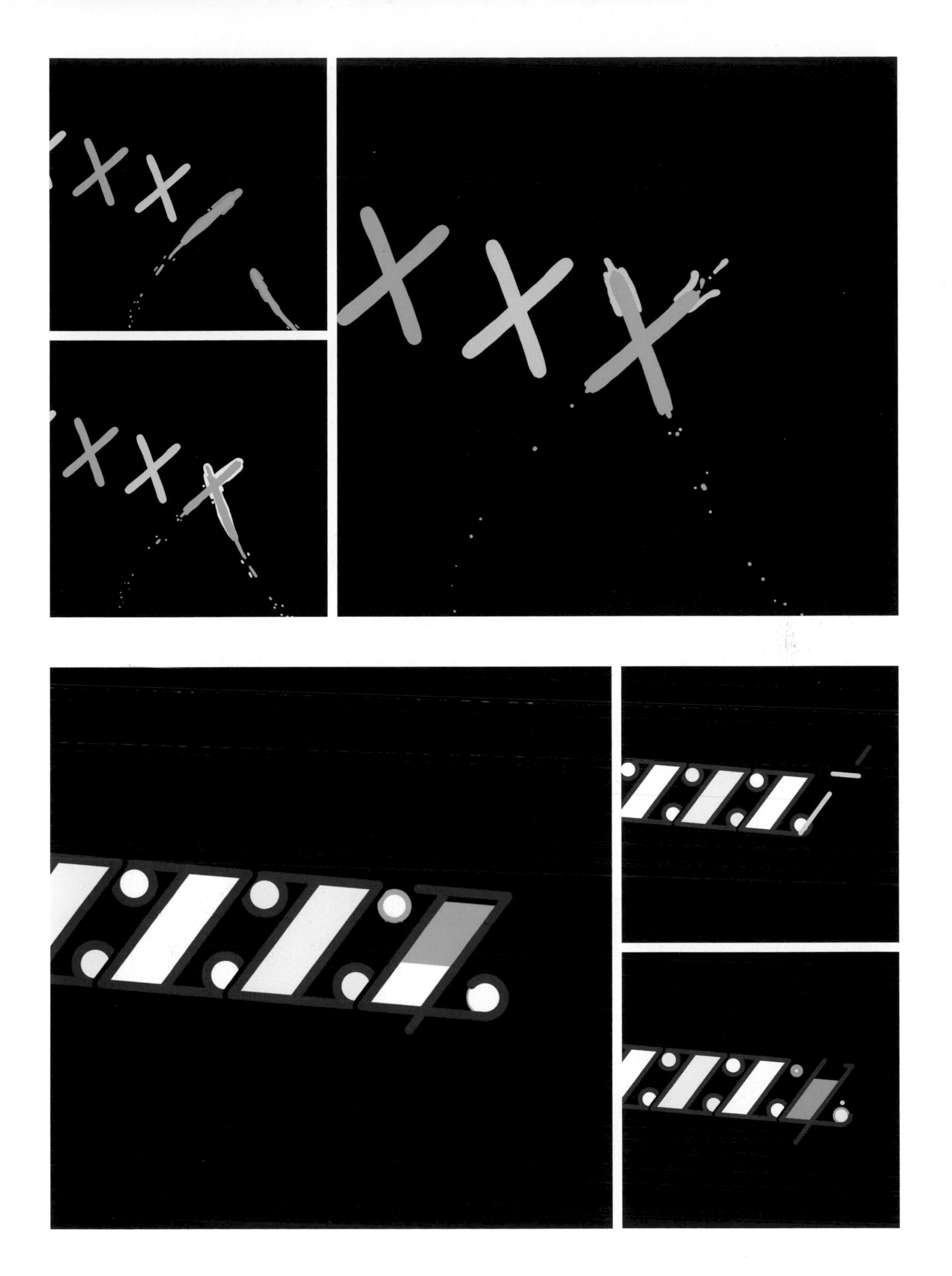

▲ TYPOGRAPHY DESIGN & ANIMATION **Emanuelle Marani** / TYPEFACE **Burstype**
▼ TYPOGRAPHY DESIGN & ANIMATION **Sara Bennett** / TYPEFACE **Jasper**

EMPTY

BY Acil Benamara, Pierre Mallart

This project is a reflection with a double meaning for vacant exhibition's spaces, normally used for advertisement. It represents the absence of publicity content, however, showing typographic animations and exhibiting emptiness.

Behance

EAA Open Days 2019 Posters

BY Mattia Marchese

Mattia Marchese was asked to create the posters for the open days of École d'Arts Appliqués de La Chaux-de-Fonds (EAA) for the second time. The typography inside these posters is flexible and responsive to various media. Both the red and white colours echo to the Swiss flag and also to the corporate identity of EAA. Overall, the posters reflect an abstract but dynamic image to the audience.

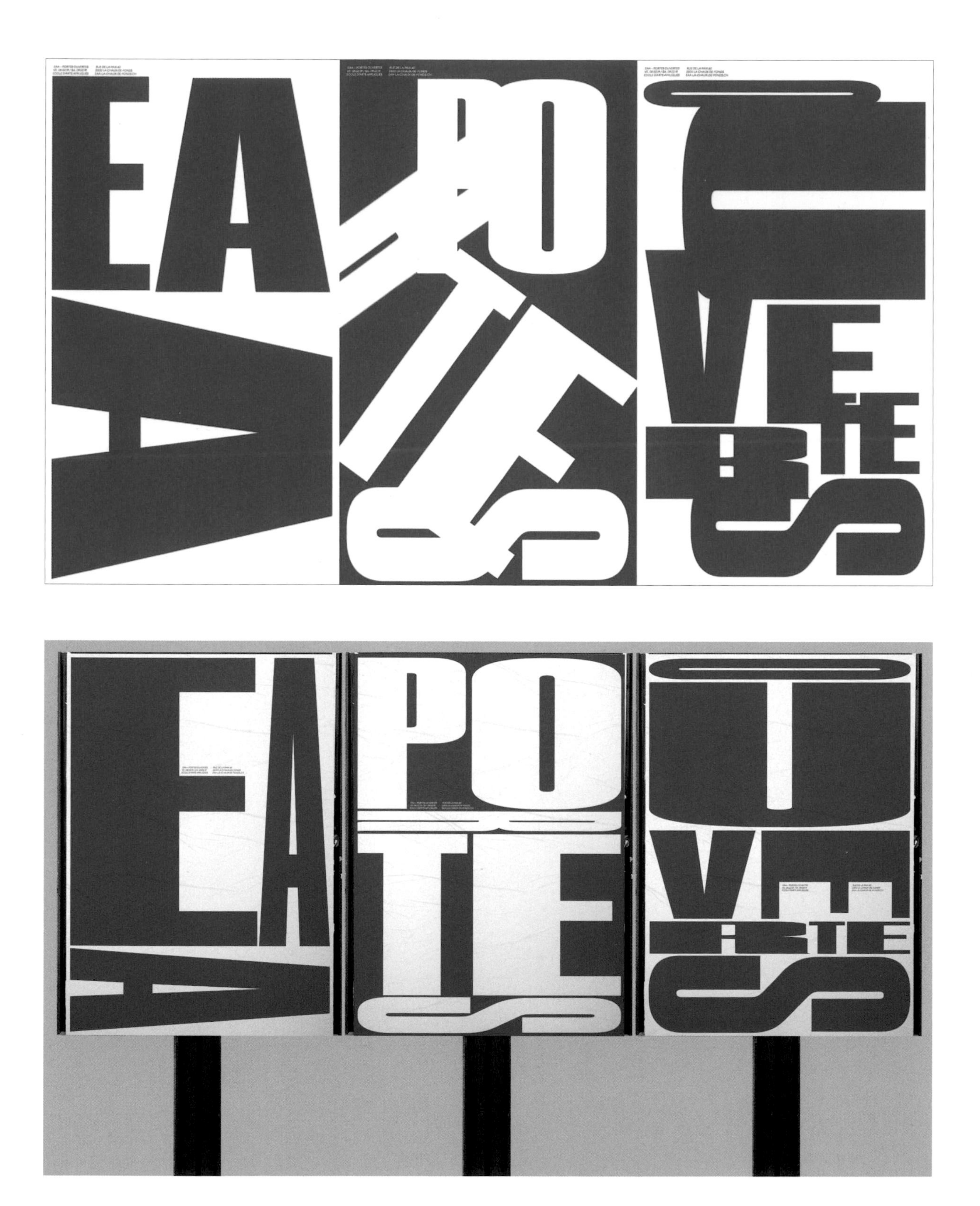

Behance
Official

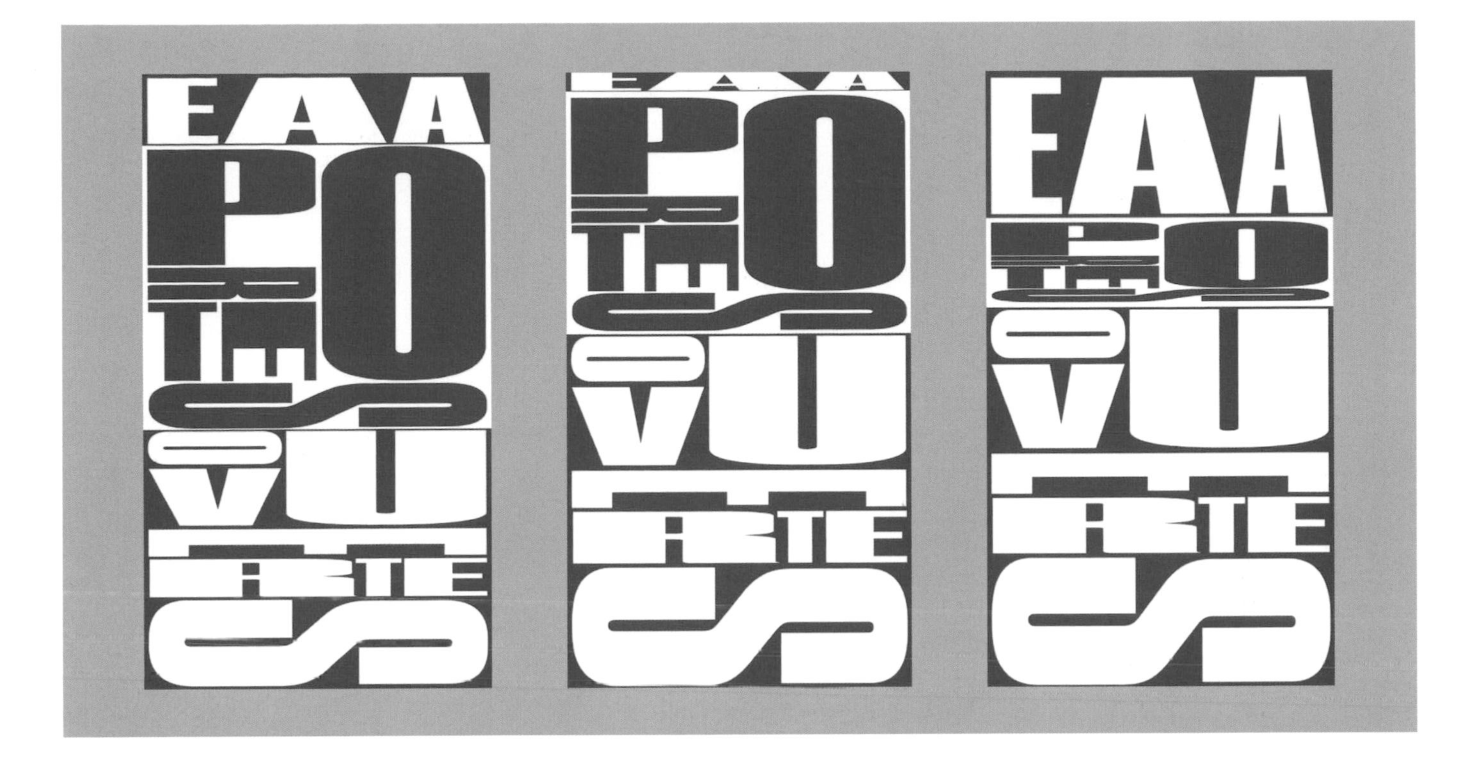
EAA
EAA
EAA

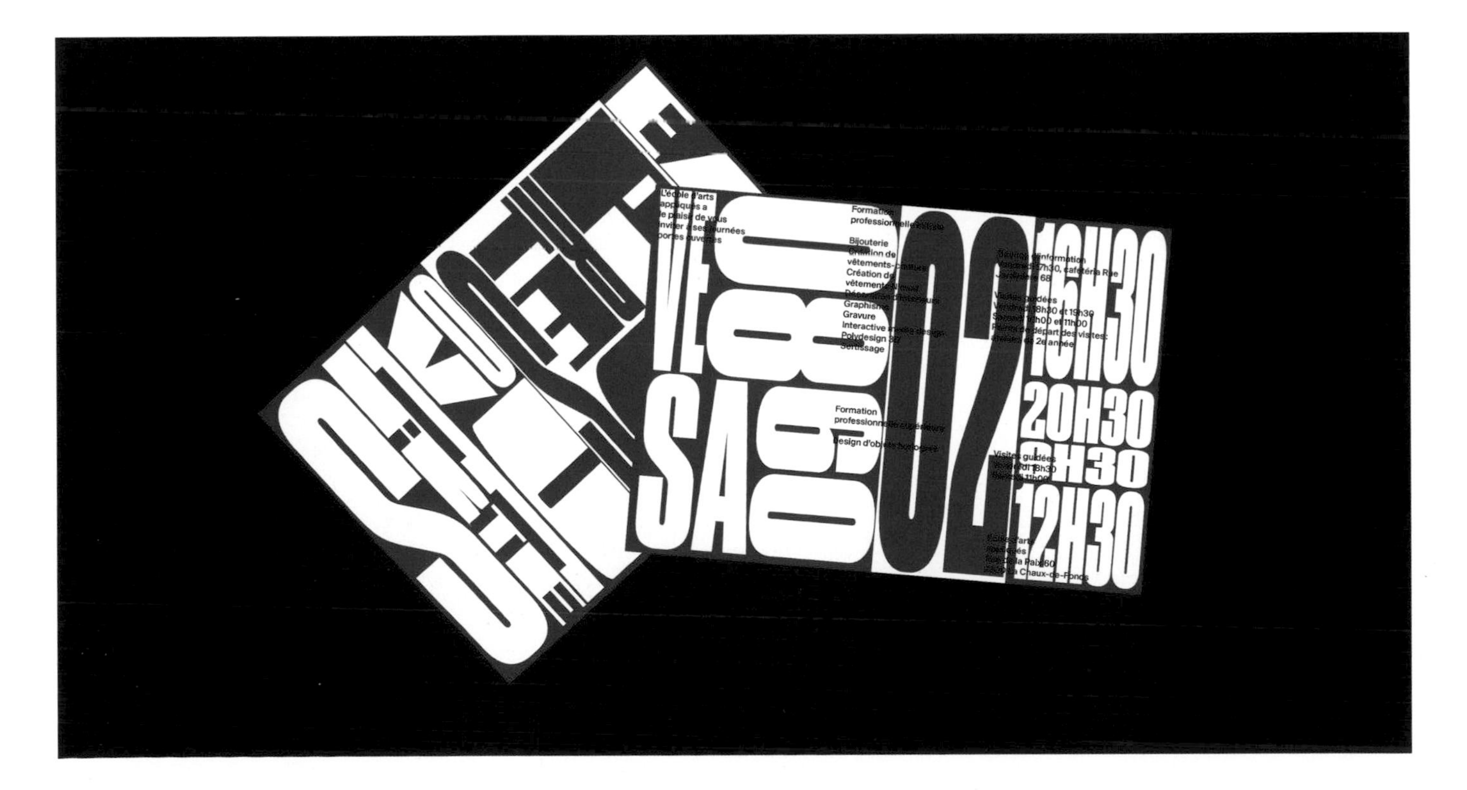
Formation
Bijouterie
Graphisme
Gravure
Sertissage
Formation
20H30
12H30
Visites guidées
La Chaux-de-Fonds

36 Days of Type

BY Sam Burton

This project was created for the 2018 edition of 36 Days of Type by Sam Burton. Sam selected a bold and vibrant colour palette with the aim of making an experimental and playful set of characters. He used both 2D and 3D design and animation techniques to create daily animated letters and numbers. Some of the characters are instantly recognisable whilst others have a more abstract appearance.

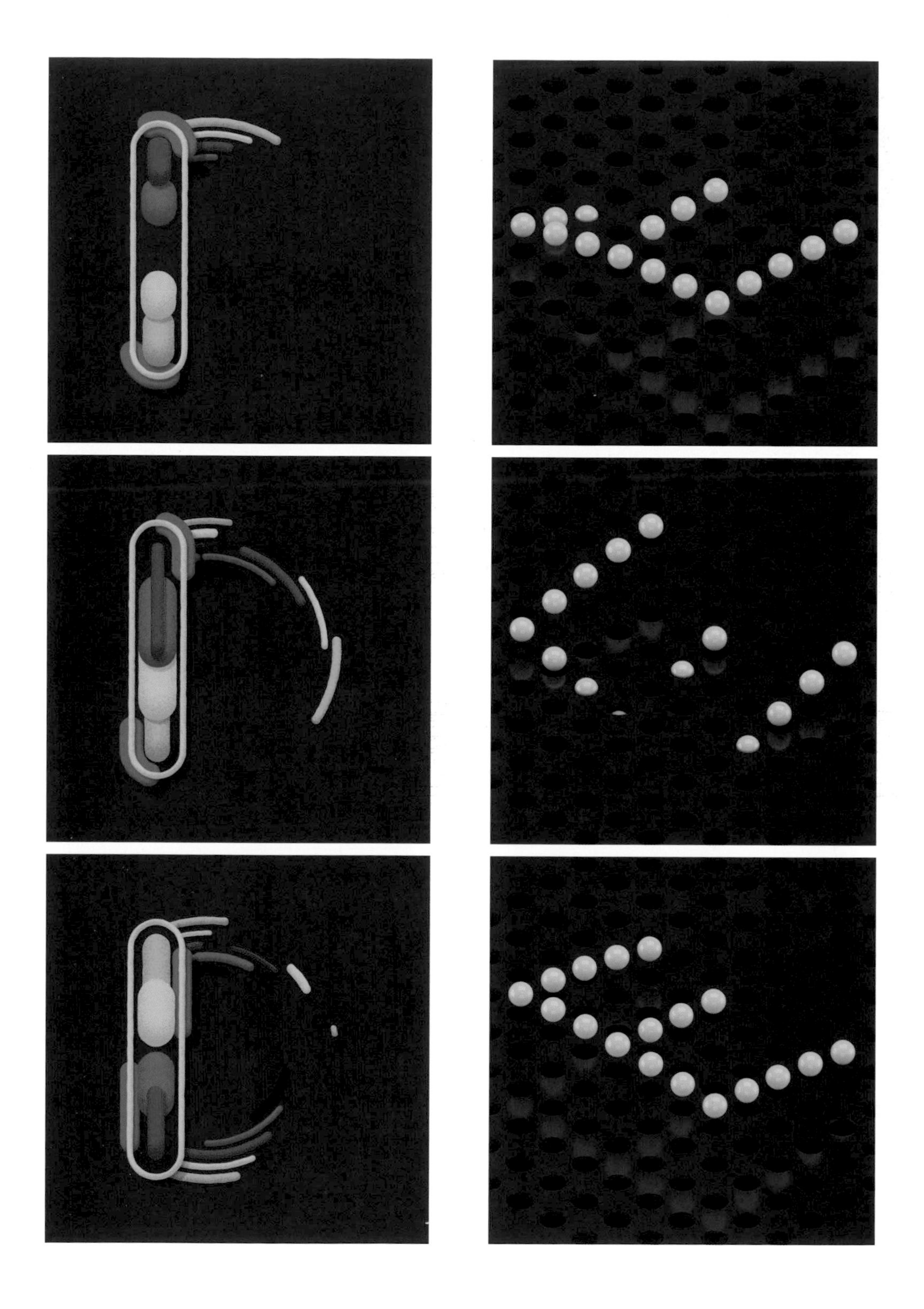

Vimeo

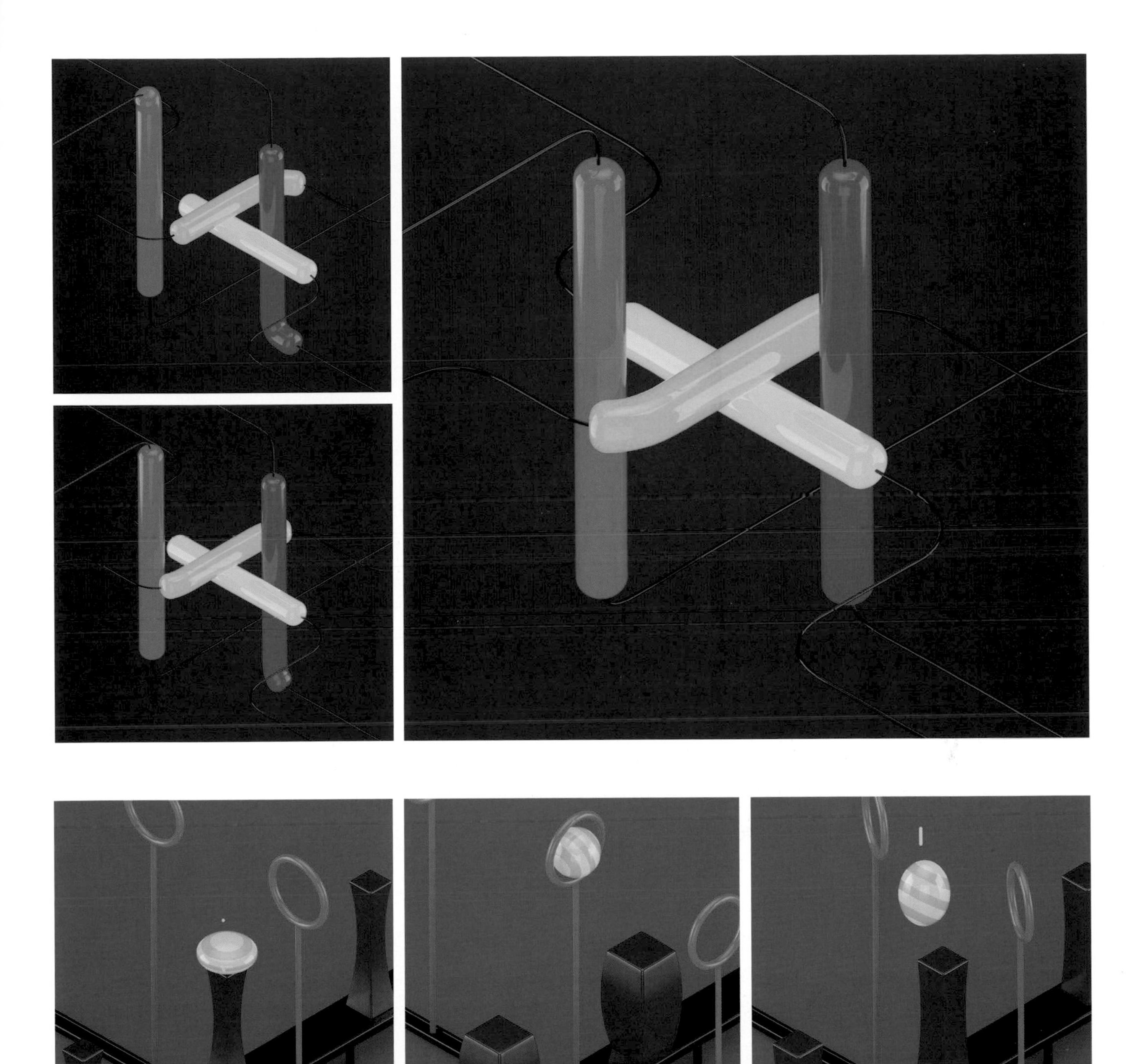

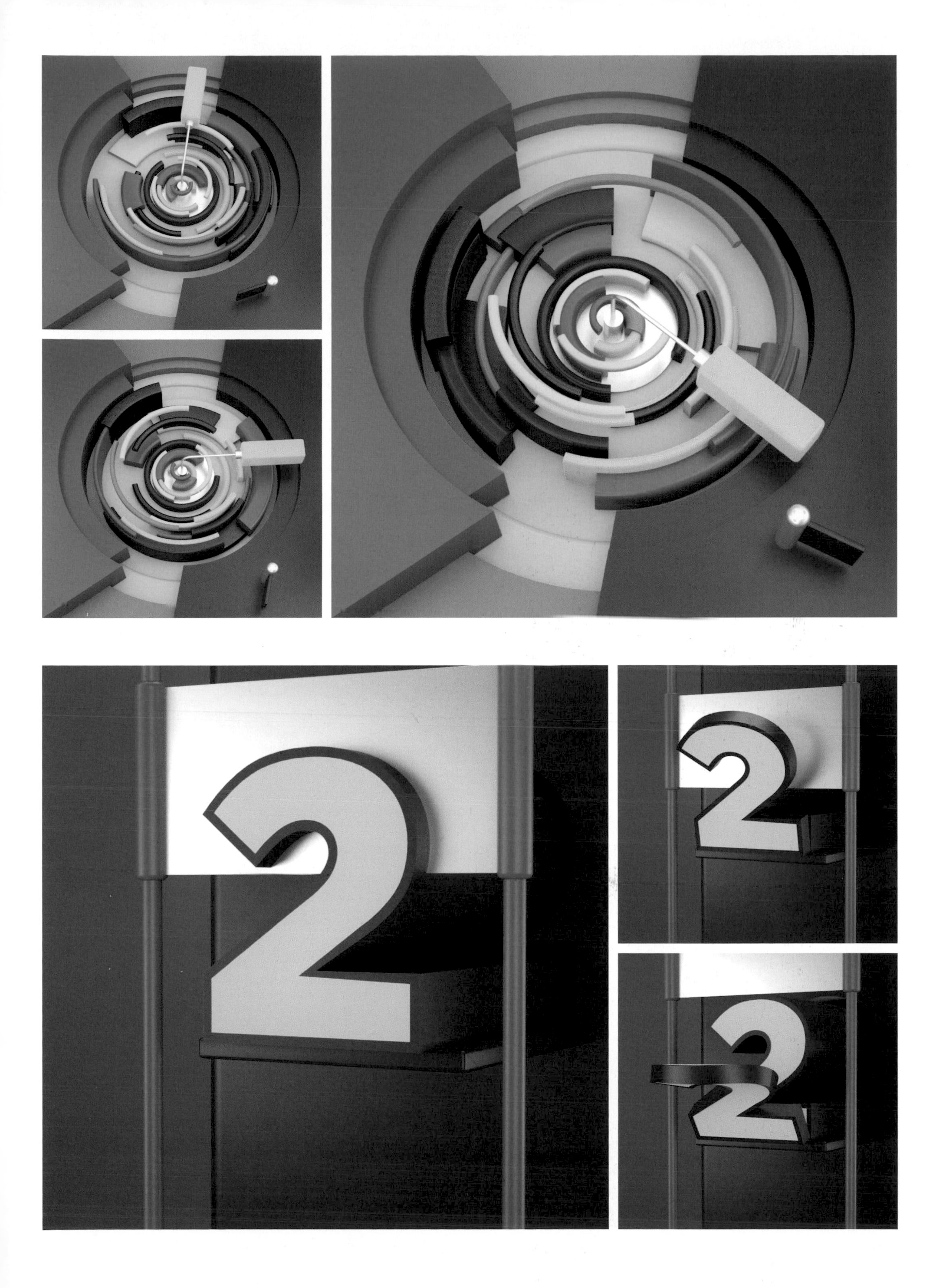

I.C.E

BY Tony S Briggs

As an experimental work, I.C.E tries to explore how typographic forms can be morphed to create the illusion of transition. The duplicate and loop slogans show the influence of 90's rave culture in the UK. And the colour palette created an illusion with a transcendent nightclub experience. The process of creation was expressive—Tony S Briggs gathered the references from the Acid House posters of the late '80s and in particular the opening credits of Gaspar Noe's *Enter the Void*. Tony combined the shape morphing, up-scale effects, and colour filters into the video by Adobe After Effects.

Dribbble

36 Days of Type

BY Sail Ho Studio

Sail Ho Studio worked on the 36 Days of Type as a self-initiated project, involving both illustration and motion design and lately released the font as free for personal use to celebrate the first year of activity. Each letter was designed in Adobe Illustrator and animated by Adobe After Effects composing and breaking up all the letters' geometrical elements.

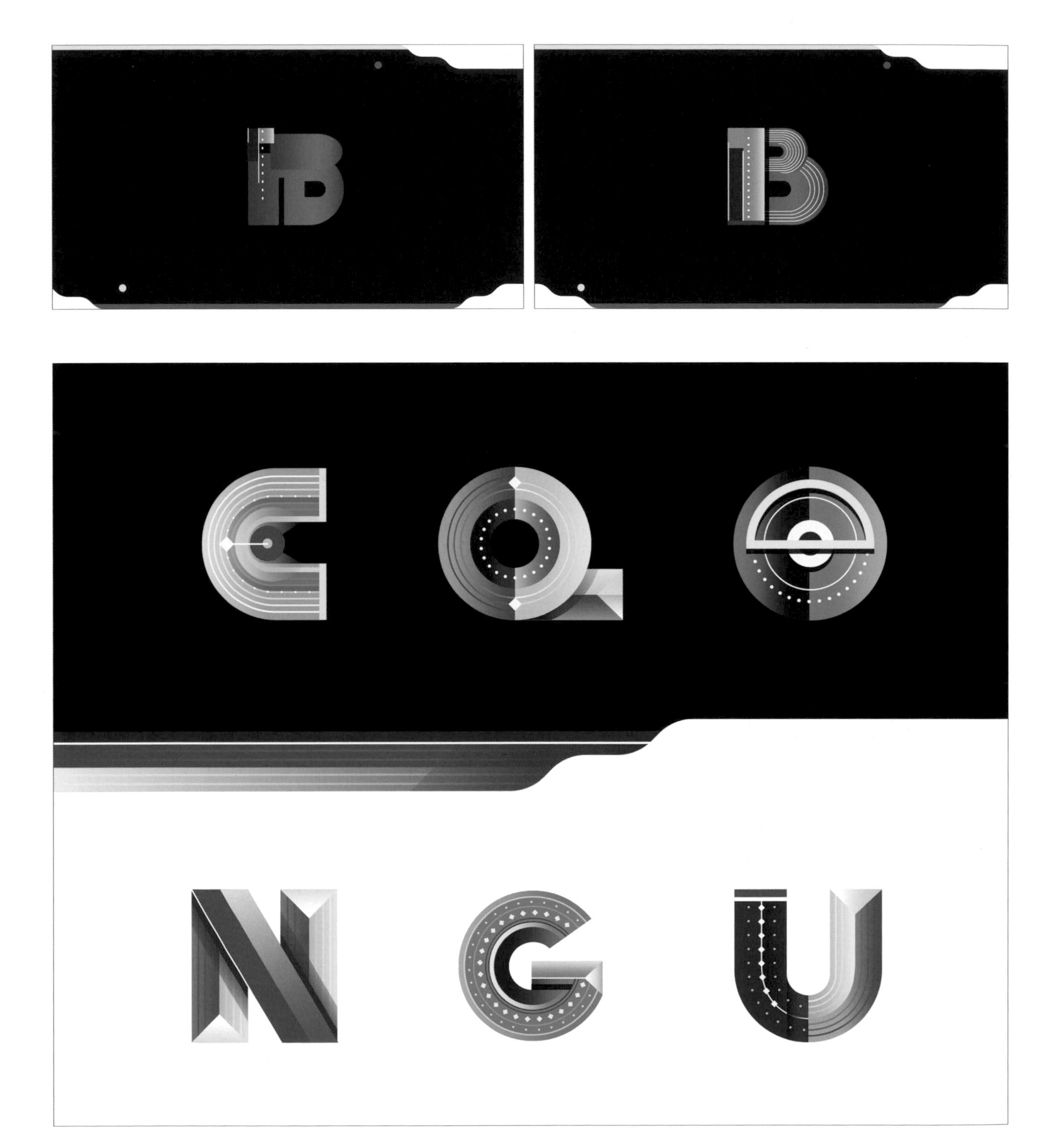

ANIMATION **Sail Ho Studio** / SOUND **Tommaso Simonetta**

Behance

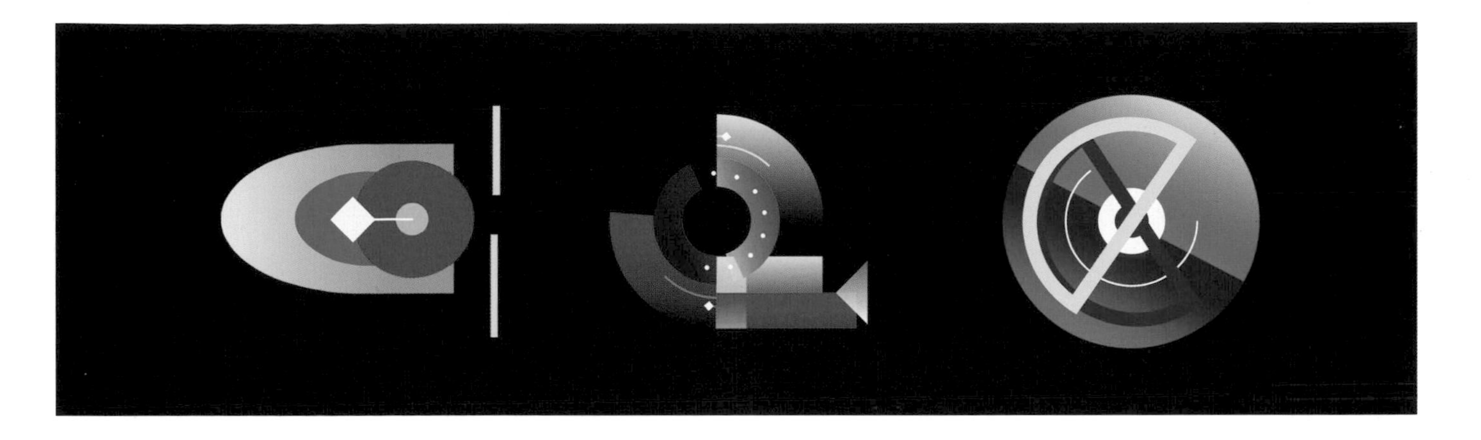

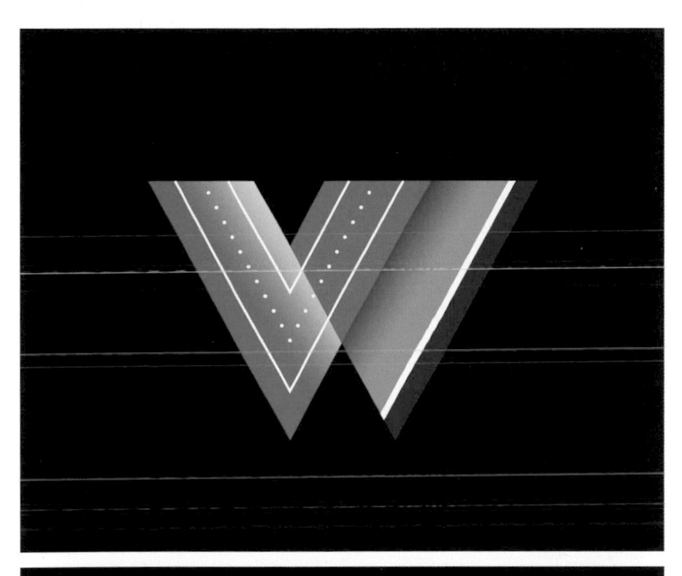

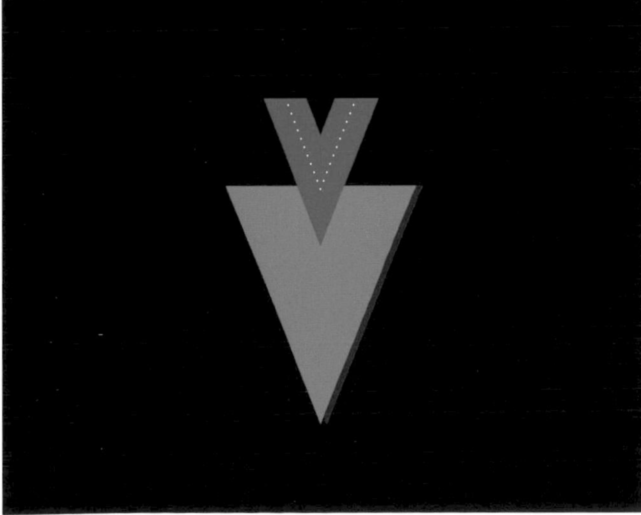

Semi Permanent Conference Titles

BY Emily Tinklenberg

Semi Permanent is a creative summit that gathers artists and designers every year. Semi Permanent's 2018 theme was "Creative Tension," or the moment right before someone finding out the perfect project solution. Using refraction and slight illegibility, this hypothetical rebranding and title sequence reimagined the tension as fluid and distorted typography that cannot be seen clearly until the very end. A combination of refracted 2D and 3D typography created an organised but fluid system.

DESIGN & ANIMATION **Emily Tinklenberg**

Vimeo Official

Type in Motion

BY Mindaugas Dudenas

Mindaugas Dudenas turned various simple words in daily life into a visual representation of their own characteristics. Mindaugas stated that the first artwork was created completely by accident. And later it became a series representing that the ideas and familiar words can be transferred by the help of typography and motion.

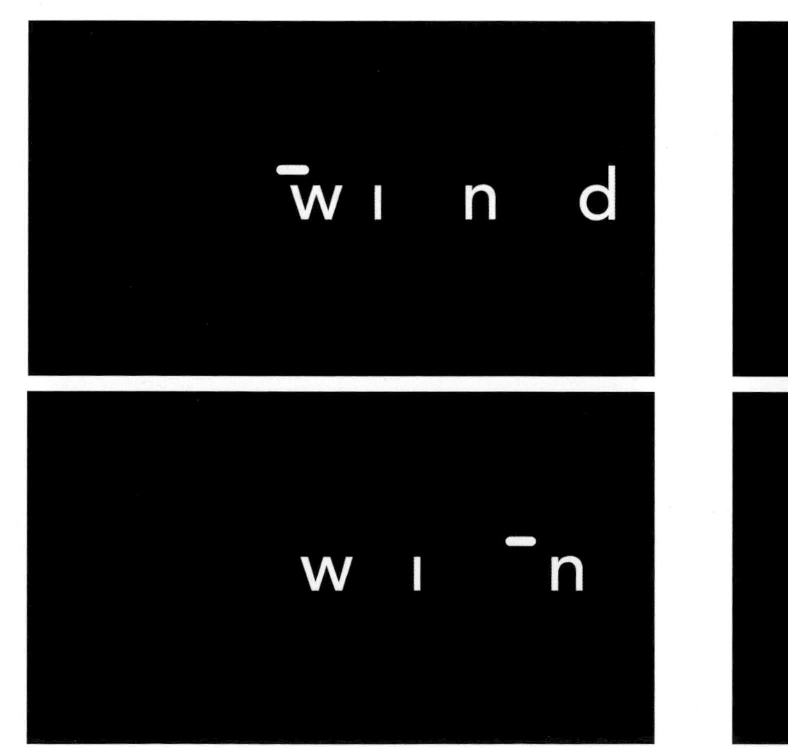

Behance

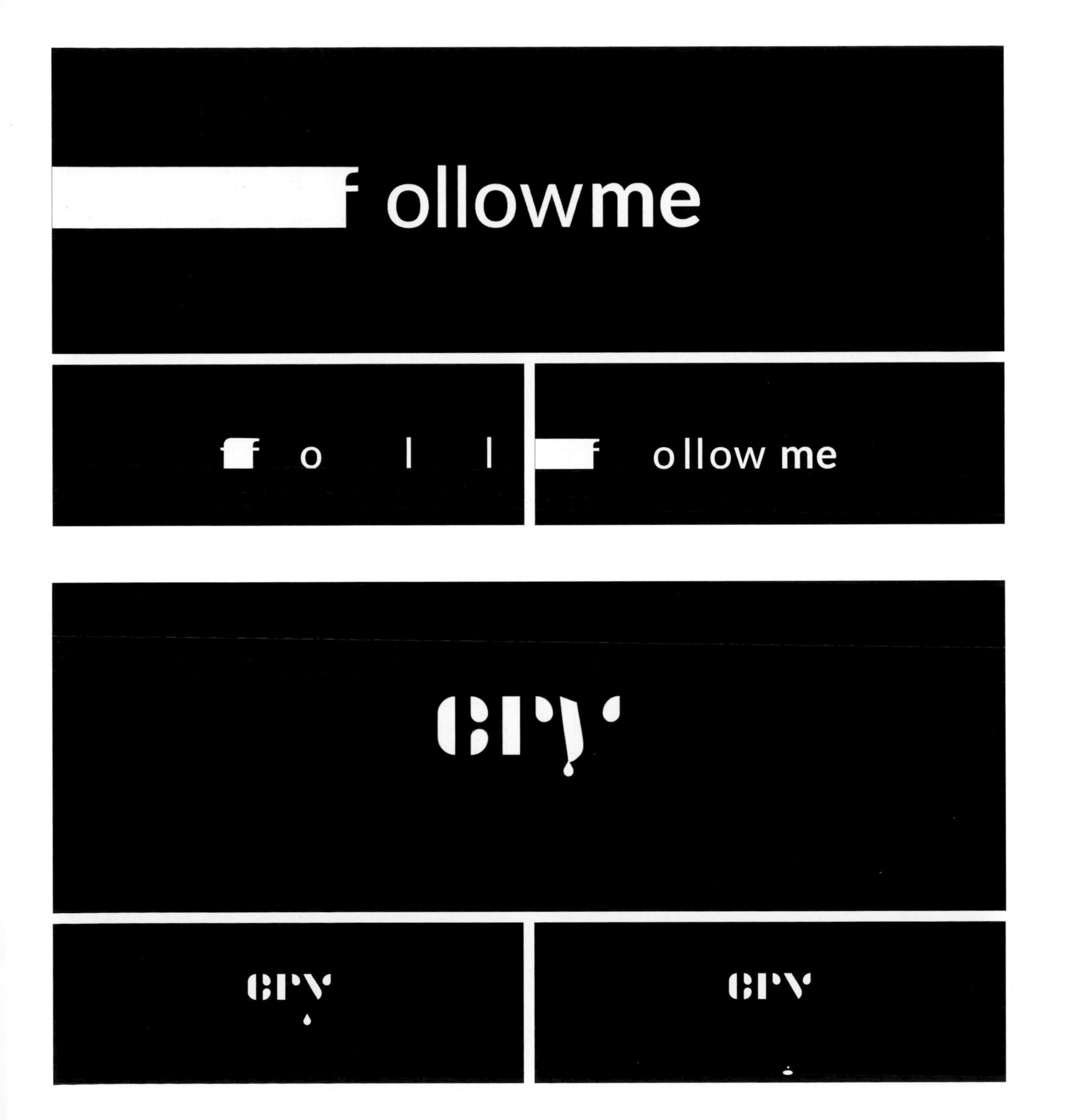
f ollowme
f o l l
f ollow me
cry
cry
cry

36 Days of Type

BY Duncan Brazzil

This project was done during Duncan Brazzil's first year of participation in 36 Days of Type. Duncan tried to explore various ways of distorting and animating the letters in order to form their own traits.

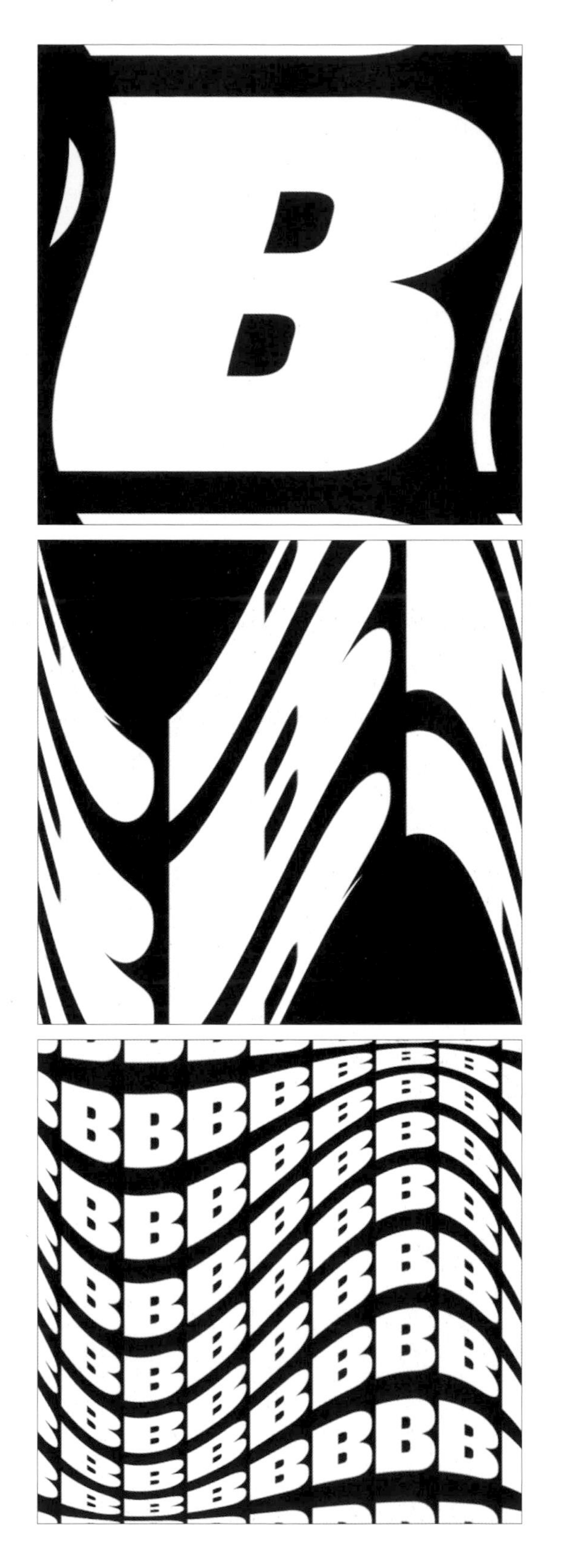

Instagram
Official

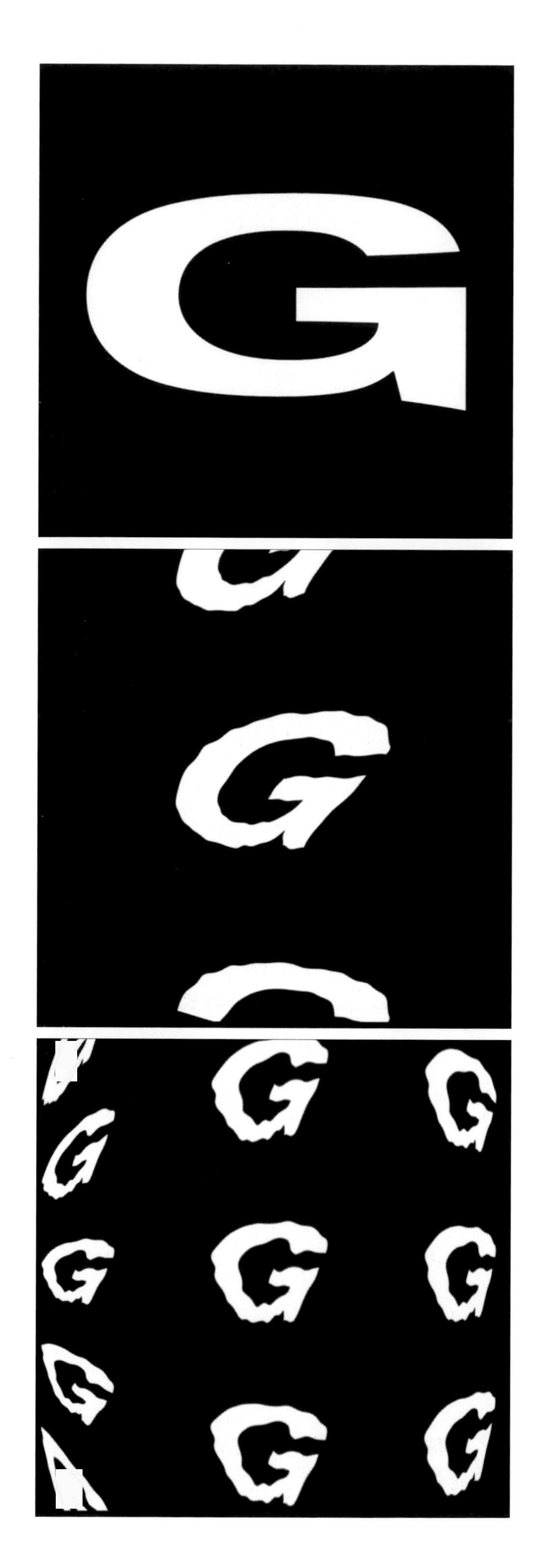

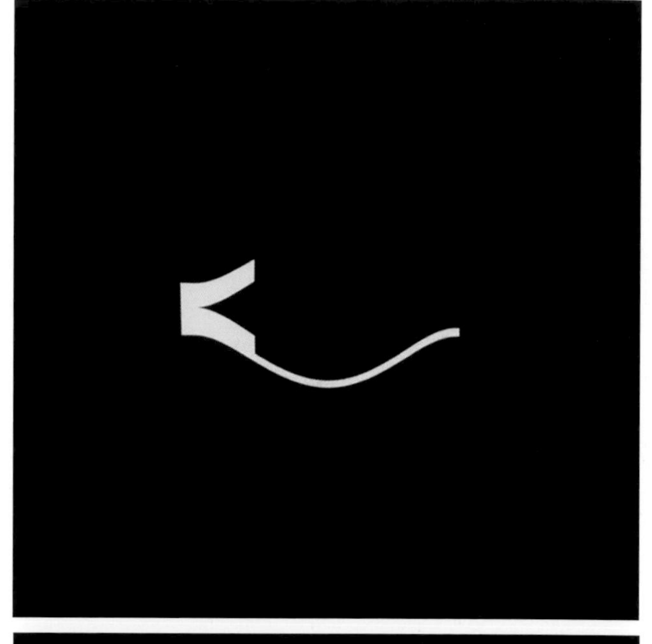

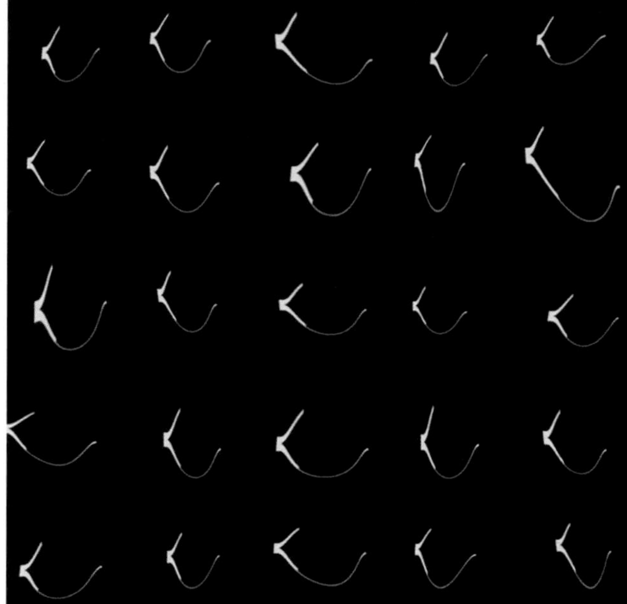

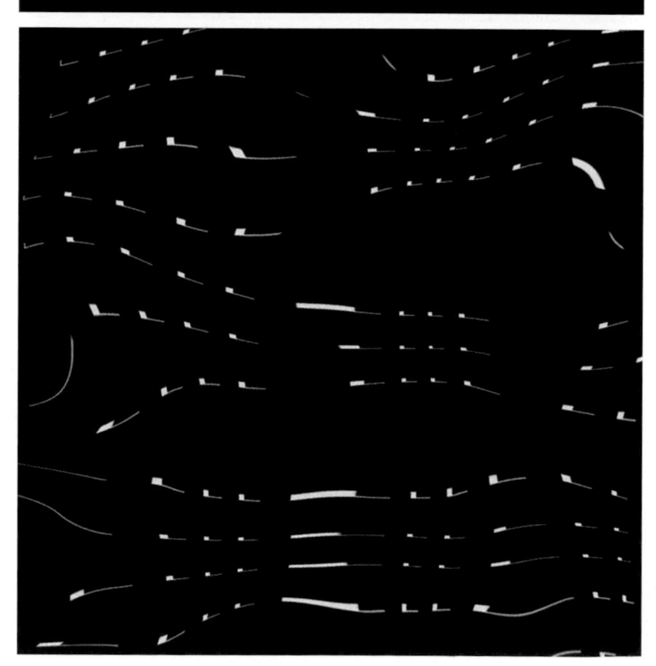

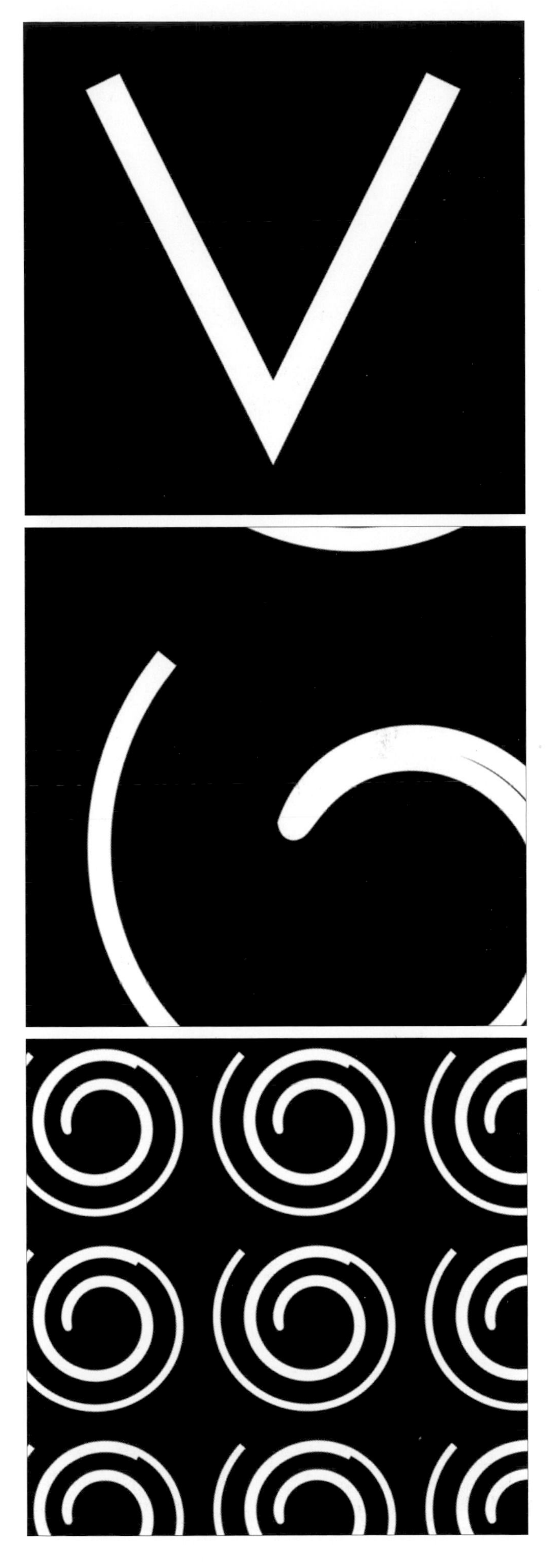

1.5 Years Creative Mornings

BY Bureau Mitte

Bureau Mitte celebrated their special event—Creative Mornings (Kreativfrühstück) that had run for over one year and a half made for the creative people in and around Frankfurt, Germany. For this special event, they created a poster series with Cinema 4D. As a key visual, the special date "1.5" added up on the posters with a numerical twist motion design.

ART DIRECTION **Anna Ranches, Helene Uhl** / MOTION DESIGN **Timo Lenzen**

INDEX

BLK OUT

blkout.agency

BLK OUT is a French design studio creating digital experiences. BLK OUT works for entertainments brands like Ubisoft, Bandai Namco, and also for different brands like Airbus Helicopters, Accor, and Whaou.

P142–143

BLOCK & TACKLE

www.blockandtackle.tv

BLOCK & TACKLE is a creative studio focusing on conceptual design and visual storytelling. Launched in New York in 2014, they have been named among The New York Times' "The Selects" group, honouring the world's top creative talent. Passionate about providing their forward-thinking clients highly original solutions from concept through delivery, BLOCK & TACKLE is built to move—with artistry that is earnest, considered, and a little mischievous.

P106–107

Bojan Milinkovic

www.bojanm.tv

Bojan Milinkovic has been working in the motion design industry for over five years. He is mainly focusing on creating 2D animations and short videos from start to finish. Searching for new ways to improve his expertise, Bojan had the chance to work in an international team with clients like Heineken, NameCheap, and GlaxoSmithKline.

P108–109

BOND

bond-agency.com

BOND is a global design agency with studios in Dubai, Helsinki, London, and Tallinn. BOND is a brand-driven creative agency with a craftsman attitude. They help new businesses and brands get started. They refresh and revolutionise existing brands for growth.

P030–031

Buck

buck.tv

Buck is a design-driven production company. They are an integrated collective of designers, artists, and storytellers, who believe in the power of collaboration—the special alchemy that only happens when working together toward a common purpose.

P112

Bureau Mitte

www.bureaumitte.de

Bureau Mitte is a Frankfurt based multi-disciplinary design and craft studio established by Anna Ranches and Helene Uhl in 2011, providing strategy, identity, and design for commercial and cultural clients.

P232

Camille Frairrot

www.steviaplease.me

Camille Frairrot (aka Stevia Please) graduated in Interactive Design of Les Gobelins's School and is interested in art direction, web design, illustration, motion design, 3D, and experimentations.

P148–149

Check it Out!

www.checkitout.studio

Check it Out! is a motion design studio from Porto, Portugal. They are into communication design, motion graphics, illustration, and character animation.

P064–065

Chen Junxian

behance.net/chenjunxian

Chen Junxian is a brand visual designer living in Shenzhen, China. Many of his works have been featured on Behance.

P180–181

CLEVER°FRANKE

www.cleverfranke.com

CLEVER°FRANKE connects people with data through powerful tools and captivation of experiences that drive change.

P152–153

Cody Cano

behance.net/CodyCano

Cody Cano is a multidisciplinary designer currently based in the Los Angeles.

P124–125

Contemple

ccccontemple.com

Contemple is a creative agency based in France owned by Antony Pradin. They provide a wide array of creative and strategic solutions for brands. They love music, design, and Internet.

P136–137

Cuchillo

cuchillo.studio

Cuchillo is a small independent creative studio located in Bilbao, Basque Country, Spain. They take special care of the bidirectional dialogue between design and development to deliver works where visual creativity and technology go hand in hand.

P114–119

Daikoku Design Institute, the Nippon Design Center, Inc.

daikoku.ndc.co.jp

Daigo Daikoku entered Nippon Design Center after graduating from the Kanazawa College of Art in 2003 and started Daikoku Design Institute in 2011. He moved to Los Angeles in 2018. He has worked with companies, educational organizations, governments, architects, and artists to create projects in various fields.

P164–165

Daniel Spatzek

www.danielspatzek.com

Daniel Spatzek is an Austrian freelance designer and developer.

P120, P126–127

democràcia estudi

www.democraciaestudio.com

democràcia estudio creates global brand experiences. democràcia estudio seeks the balance between design and communication to achieve unique solutions, working in a complementary and transversal way to offer global results. Ideas that were born between people create experiences for the world.

P182–183

Duncan Brazzil

duncanbrazzil.com

Duncan Brazzil (aka Design By Duncan) is an American/British graphic designer specialising in motion, typography, digital, print, and brand identity residing in Edinburgh, Scotland. Currently residing in Portland, Oregon, he is exploring the subject of kinetic typography with highlighted projects "36 Days of Type" and "Kinetic Typography Experiments."

P156–159, P228–231

Éléonore Sabaté

www.eleonoresabate.com

Éléonore Sabaté is a graphic designer and art director from Paris, France.

P016–017

Emily Tinklenberg

www.emilytink.com

Emily Tinklenberg is a designer and animator from Detroit. She graduated from College for Creative Studies in 2019 with a degree in Communication Design and has a focus on motion graphics and animation. With an emphasis on engaging storytelling, she strives to create weird, playful, and subversive works that can make people happy for just one moment of their days.

P224–225

estúdio arco

estudioarco.com.br

estúdio arco is a multidisciplinary design studio based in São Paulo. The studio works through print and digital, connecting functional with experimental approaches and aiming to develop projects consistent in their history, functionality, and visual attributes, communicating in a relevant and genuine way.

P178–179

FIELD

field.io

FIELD is a professional creative studio in London and Berlin. They create powerful new formats of brand communication using art and technology—expressing values, ambitions, and the complexity of processes in universal and emotional artworks.

P090–091

Flora de Carvalho

floradecarvalho.com

Flora de Carvalho likes to work where graphic and type design meet, creating custom type and lettering to display text in joyous and smart ways. Since 2018 she has worked under Passeio—a studio shared with illustrator Dominique Kronemberger.

P178–179

Frame

frame.dk

At the intersection of design, animation, and film, Frame directs, produces, and collaborates to create the top-shelf design, animation, and motion graphics for screens of all sizes.

P042–043

Fromsquare Studio

fromsquare.com

Fromsquare is a Poznań based graphic design studio founded by graphic designers Alicja Piotrowska and Jakub Piechota. They specialise in animation and branding. Their main goal is to create a coherent and effective communication, which reflects in the creation of dynamic brands and visual identification systems.

P032–033

Gimmick Studio

gimmickstudio.tv

Gimmick Studio is an award-winning motion design studio based in Montreal, Canada. Specialising in 2D and 3D animation, typographic animation, and character animation, Gimmick Studio works with a broad range of clients in the advertising, television, and multimedia industry.

P014–015

Glare

glare.la

Glare is a future-forward creative studio in Los Angeles run by the digital artist duo Anthony Gargasz and Rafael Ramirez.

P020–021

Holke79

www.holke79.com

Holke79 (aka Borja Holke) is a motion graphic designer. With more than ten years working in this field, he loves his work and likes a satisfied client more than anything.

P184–185

Hunk Xing

behance.net/ihunk

Hunk Xing is a graphic designer based in Beijing, China. His works have been included in various publications, such as *Asia Pacific Design*, *New 3D Effects in Graphic Design*, and *Asian Typography*. He is also the winner of the 2017 and 2018 Zcool Excellence Award.

P196–197

INLANDSTUDIO

inlandstudio.tv

INLANDSTUDIO is a design and animation company based in Buenos Aires producing branding and commercial with a powerful 3D, a careful and detailed motion, and a bold and unique creative direction.

P088–089

Javier Miranda Nieto

behance.net/Javimir

Javier Miranda Nieto is a minimal and abstract motion designer from Neuquén, Argentina.

P044–045

Justin Au

www.justin5au.com

Justin Au is a multidisciplinary designer, art director, and animator with a focus on dynamic brand systems, interactions, and experiences. Currently, he is working as a senior designer at Droga5 in New York.

P170–171

Kid Pixel

kidpixel.fr

Kid Pixel (aka Camille Moine) is a freelance motion graphics designer based in France. Kid Pixel combines primitive emotion and passion for the digital arts. Kid Pixel likes to say that he is the "missing pixel" to the communication.

P200, P202–203

Kurppa Hosk

kurppahosk.com

Kurppa Hosk is a diverse team of design thinkers and design doers. They collaborate with brave organisations—from early-stage startups to some of the world's most admired companies. They use business, brand, and experience design to inform and inspire meaningful progression.

P048–049

Lumiko

www.lumiko.com.au

Lumiko is the portfolio of Sydney based director and designer Chris Thompson. With years of experience in areas both creative and technical Chris continues to explore the lands between moving image, design, and technology.

P038–039

MamboMambo

mambomambo.ca

MamboMambo is a young brand and interactive agency from Quebec, Canada.

P078–079

Manuel Martin

www.manuelmartin.design

Manuel Martin was born in Venezuela and is an art director, designer, and animator.

P082–083, P094

Manuel Rovira

behance.net/ManuelRovira

Manuel Rovira is a web & UI/UX designer and co-founder of Lusaxweb and Vuesax.

P122–123

Mattia Marchese

mattiamarchese.com

Mattia Marchese is a young graphic designer with a strong focus on typography and a deep influence from the Swiss Style in the 1950s.

P212–213

Megan Palero

minority.work

Megan Palero is a motion designer based in Davao City, Philippines who specialises in broadcast and design research. He introduces research methods in engaging human stories through animation and any other forms of moving experiences. Designing with types is a vital phase in his process of producing work for commercial or social campaigns and brand identities. And he also founded a visual journal called *Urban Type of Davao*.

P104–105

MEMOMA Estudio

memoma.tv

MEMOMA Estudio is a creative team passionate about motion design. They are the innovators ready to jump from trend to trend. They seek creative solutions, synthesis, and metamorphosis of ideas that meet the needs of emerging markets. They are a team of creative multifaceted baroques, lovers of unexpected results.

P034–035

Mindaugas Dudenas

dudenas.net

Mindaugas Dudenas is a designer from Vilnius, Lithuania.

P226–227

Moodagency

moodagency.pl

Moodagency talks and listens. And afterwards, they think, write, design, and produce. For Moodagency, the meaning and the image are equally important. They combine aesthetics with simplicity. They want people in the team to understand each other. They like teamwork and good mood in the studio. They are mood.

P168–169

Mother Design

www.motherdesign.com

Mother Design is an independent branding and design studio founded in 2006. They specialise in non-specialising, with a diverse portfolio of works and clients, brought together by a shared desire to make the world a better place through design.

P024–025

NŌBL

www.nobl.tv

NŌBL is a creative studio specialising in art direction and design. They are passionate about creating meaning by using movement in all their forms.

P040–041, P046–047

Nuno Leites

nunoleites.com

Nuno Leites is a freelance motion designer based in Porto, Portugal.

P198–199

OddOne

www.oddone.nl

OddOne is a visual storyteller that develops creative solutions for advertising agencies, production houses, and private businesses. By designing visual languages with motion in mind, OddOne supplies their clients with visual narratives that are dynamic and unique.

P092–093

Oh Yeah Studio

www.ohyeahstudio.no

Oh Yeah Studio was founded by Hans Christian Øren, a Norwegian designer based in Oslo.

P018–019

Papanapa

papanapa.com

Papanapa is a graphic and motion design studio based in São Paulo. Their goal is to create meaningful and aesthetically attractive experiences for clients.

P080–081

Pes Motion Studio

www.peslab.com.ar

Pes is a motion studio specialising in integrating fresh ideas, creativity, and experience. From offices in Buenos Aires and Chicago, Pes offers a collaborative multicultural perspective to their clients in the advertising, Internet, broadcast, film, and entertainment industries.

P068–069, P076–077

Pierre Mallart

pierremallart.com

Pierre Mallart is a freelance art director and motion designer based in Paris.

P210–211

Plau Design

plau.co

Plau Design makes type pop for people and the brands they love. They teach and speak about type to everyone with a slight interest in this wonderful world of letters. Their works have been used and seen by millions of people in daily life.

P178–179

Ranger & Fox

rangerandfox.tv

Ranger & Fox is a creative studio that specialises in discovery, strategy, and visual communication, elevating forward-thinking brands like Microsoft, UFC, Paramount Pictures, and HP through thoughtful and effective motion design.

P054–055, P096–097

Red Collar

redcollar.digital

Red Collar is the Agency of the Year announced by international web design and development award platform CSS Design Awards. They create impressive digital things that help brands reach both the minds and hearts of people.

P133

Rémi Vincent

www.remi-vincent.com

Rémi Vincent is a freelance art director and motion designer based in Montreal, Canada.

P070–071

République Studio

www.republique.studio

République Studio is an award-winning agency. They work with local and international clients across a diverse range of platforms and disciplines. Typography is one of the key elements of their works for designing brand identities, signage, publications, magazines, posters, and websites. Their strength is in versatile and modern design, often inspired by the zeitgeist.

P022–023

Rethink

rethinkcanada.com

Rethink is an independent creative agency here to help you rethink.

P070–071

Roberto Warner

robertowarner.info

Roberto Warner is a communication designer, currently working in San Francisco.

P194–195

Romain Granai

romaingranai.be

Romain Granai is a creative front-end developer and graphic designer from Liege, Belgium, currently living and working in London, UK. He is particularly interested in various printing processes such as silkscreen printing, experimenting with typography and minimal website design.

P138–139

Safari Riot

noise.safaririot.com

Safari Riot is a multi-disciplinary creative studio exploring the future of media. Their :Noise Division focuses on sound and music, while :XP Division develops interactive XR (Extended Reality).

P144–145

Sail Ho Studio

www.sailhostudio.com

Sail Ho Studio is an illustration and motion design collective, founded by five designers living across Europe—Mirko Càmia (illustrator), Matteo Goi (motion designer), Davide Mazzuchin (illustrator), Daniele Simonelli (illustrator), and Chiara Vercesi (illustrator) .

P220–221

Sam Burton

www.sambmotion.com

Sam Burton is a freelance motion designer and animator. For the last ten years, he has worked across a large variety of projects including animated commercials, broadcast design, music videos, and pretty much everything in between.

P214–217

Sebastien Camden

camden.work

Sebastien Camden is a motion designer and director living and working in Montreal, Canada on various creative endeavours, such as explainer videos, promotional contents, advertisements, short films, and music videos.

P062–063, P095, P102–103, P110–111

Sonya Nechkina

behance.net/sonyanechkina

Sonya Nechkina is a motion designer from St. Petersburg, Russia.

P052–053

Souffl

souffl.com

Souffl is a design and innovation company helping its clients to explore, develop, and build the future of their business. Souffl was founded in 2014 by three partners—Nicolas Baumgartner, Arnaud Carrette, and Fabien Fumeron—around a shared vision of designing a more humane and conscious future.

P146–147

Studio BYTS

vimeo.com/byts

BYTS (Bright Young Things) is a Seoul based studio specialising in music videos, commercials, motion graphics, and short films.

P060–061, P066–067

Studio Dumbar

studiodumbar.com

Studio Dumbar (Part of Dept) is an international agency with a Dutch heritage, specialising in visual branding, founded by Gert Dumbar in 1977. Studio Dumbar attracts talented individuals from around the world. Their portfolio is equally diverse, encompassing works for a variety of clients both large and small—from business and government to cultural and non-profit.

P134–135

Styleforces

youtube.com/channel/UClpXxRZ6Ep_8Y160do-GJpA

Styleforces is an art direction and design studio. Based in Warsaw, they make templates by Adobe After Effects and all kinds of clever stuff related to 2D and 3D graphics.

P176–177

Syddharth Mate

syddharth.com

Syddharth Mate is a Master's student enrolled in the Graphic Design program at the National Institute of Design, Ahmedabad, India.

P186–189

Timothée Roussilhe

timroussilhe.com

Timothée Roussilhe is a creative developer and designer. He enjoys creating beautiful and thoughtful experiences. He likes to mix codes into surprising visuals and pleasing interactions.

P140–141

Ting-An Ho

tinganho.info

Ting-An Ho, born in 1991, is a graphic designer and art director based in Taiwan, China. In 2011 he started putting himself forward and volunteered to redesign the identity of his college, and began as an art director. As an award-winning designer, he was recommended to become one of the most influential young designers in Asia, receiving numerous recognitions including iF Design Awards.

P201

Tony S Briggs

www.tonysbriggs.co.uk

Tony S Briggs is an artist and designer based in London, UK. As a designer, he has worked with companies in the fashion, television, and retail industries. As an artist, his practice is at the intersection of motion, code, and graphic design. He takes his influence from UK club culture and underground movements.

P218–219

Transwhite Studio

www.transwhite.cn

Transwhite Studio was founded in 2011 in Hangzhou, China. It is a multifaceted design studio featuring experimental design as well as a place for communication. With a primary focus on graphic design, the studio expands their role to art exhibitions, social events, experimentations, and cross-disciplinary collaborations.

P222–223

True Form

trueform.nl

True Form is a small motion design studio based in the Netherlands. Since every brand, client, or person has a unique story, True Form with their passion animates and visualises their stories the way they are meant to be.

P056–057

Tubik Studio

tubikstudio.com

Tubik Studio has been on the market since 2013. They started from outsourcing UI/UX design and have grown the diversity of services year by year. And now Tubik Studio is a full-stack digital agency with all the specialists for the efficient and creative process from scratch.

P130–131

Vaidehi Vartak

behance.net/vaidehivartak

Vaidehi Vartak is an Indian designer specialising in UX design. Her interests lie in crafting experiences for the digital and physical world through explorative use of type, motion, and colours.

P172–173

Vince Hegedűs

www.instagram.com/vince_hegedus/

Vince Hegedűs is a multidisciplinary designer based in Budapest.

P162–163

Vincent Dumond

www.dvmg.tv

Vincent Dumond is an art director and motion designer based in Paris. Meanwhile, Vincent is a passionate artist. Precise and obsessed with smooth workflows, he is the force for bringing forward proposals.

P084–087

Vincent Raineri

www.vincentraineri.com

Vincent Raineri is an art director and motion designer based in Montreal, Canada.

P070–071

Vincenzo Marchese Ragona

vmragona.com

Vincenzo Marchese Ragona is a London based graphic designer currently enrolled in a Graphic Design course at Ravensbourne University. As an artist, he is always keen to learn something new and explore new facets of design to improve his skills even further.

P190–193

Whitelight Motion

whitelightmotion.tv

Whitelight Motion was founded by Rex Hon. As an integrated agency, Whitelight Motion offers a bespoke service for each client in all aspects of branding communication, ceremony image design, digital advertising, and interactive image installation.

P074–075

Wladyslaw Lutz

vimeo.com/blaqmatrix

Wladyslaw Lutz is a motion graphics designer from Krakow, Poland.

P098–099

Xabi Mendibe

xabimendibe.com

Xabi Mendibe is a multidisciplinary designer based in Barcelona focusing on graphic design, motion graphics, and art direction. As a freelancer, he creates most of his works for design studios and audiovisual production companies but always leaving a part to the experimentation in personal projects.

P036–037

Xtian Miller

xtian.design

Xtian Miller is an award-winning designer and renowned visual artist based in Detroit.

P132

Zoë Barber

zoe-barber.com

Zoë Barber is a designer and copywriter based in Sydney, Australia, working across branding, print, digital, and advertising. Her designs have been recognised by Awwwards, Webby, and FWA, and published locally and overseas.

P128–129

ACKNOWLEDGEMENTS

We would like to express our gratitude to all of the designers and companies for their generous contribution of images, ideas, and concepts. We are also very grateful to many other people whose names do not appear in the credits, but who made specific contributions and provided support. Without them, the successful compilation of this book would not have been possible. Special thanks to all of the contributors for sharing their innovation and creativity with all of our readers around the world.